T0259711

# SpringerBriefs in Applied Sciences and Technology

SpringerBriefs present concise summaries of cutting-edge research and practical applications across a wide spectrum of fields. Featuring compact volumes of 50 to 125 pages, the series covers a range of content from professional to academic.

Typical publications can be:

- A timely report of state-of-the art methods
- An introduction to or a manual for the application of mathematical or computer techniques
- A bridge between new research results, as published in journal articles
- A snapshot of a hot or emerging topic
- An in-depth case study
- A presentation of core concepts that students must understand in order to make independent contributions

SpringerBriefs are characterized by fast, global electronic dissemination, standard publishing contracts, standardized manuscript preparation and formatting guidelines, and expedited production schedules.

On the one hand, **SpringerBriefs in Applied Sciences and Technology** are devoted to the publication of fundamentals and applications within the different classical engineering disciplines as well as in interdisciplinary fields that recently emerged between these areas. On the other hand, as the boundary separating fundamental research and applied technology is more and more dissolving, this series is particularly open to trans-disciplinary topics between fundamental science and engineering.

Indexed by EI-Compendex, SCOPUS and Springerlink.

More information about this series at http://www.springer.com/series/8884

Igor Meglinski · Liliya Trifonyuk ·
Victor Bachinsky · Oleh Vanchulyak ·
Boris Bodnar · Maxim Sidor ·
Olexander Dubolazov · Alexander Ushenko ·
Yurii Ushenko · Irina V. Soltys · Alexander Bykov ·
Benjamin Hogan · Tatiana Novikova

# Shedding the Polarized Light on Biological Tissues

Springer

*Authors*
See next page

ISSN 2191-530X ISSN 2191-5318 (electronic)
SpringerBriefs in Applied Sciences and Technology
ISBN 978-981-10-4046-7 ISBN 978-981-10-4047-4 (eBook)
https://doi.org/10.1007/978-981-10-4047-4



This Springer imprint is published by the registered company Springer Nature Singapore Pte Ltd.
The registered company address is: 152 Beach Road, #21-01/04 Gateway East, Singapore 189721, Singapore

Igor Meglinski
College of Engineering and Physical Sciences
Aston University
Birmingham, UK

Victor Bachinsky
Department of Forensic Medicine
Bukovinian State Medical University
Chernivtsi, Ukraine

Boris Bodnar
Department of Forensic Medicine
Bukovinian State Medical University
Chernivtsi, Ukraine

Olexander Dubolazov
Yuriy Fedkovych Chernivtsi National University
Chernivtsi, Ukraine

Yurii Ushenko
Yuriy Fedkovych Chernivtsi National University
Chernivtsi, Ukraine

Alexander Bykov
Opto-Electronics and Measurement Techniques Laboratory
University of Oulu
Oulu, Finland

Tatiana Novikova
Laboratoire de Physique des Interfaces et Couches Minces (LPICM), Ecole polytechnique, IP Paris
CNRS
Paris, France

Liliya Trifonyuk
Department of Urology
Rivne Regional Hospital
Rivne, Ukraine

Oleh Vanchulyak
Department of Forensic Medicine
Bukovinian State Medical University
Chernivtsi, Ukraine

Maxim Sidor
Yuriy Fedkovych Chernivtsi National University
Chernivtsi, Ukraine

Alexander Ushenko
Yuriy Fedkovych Chernivtsi National University
Chernivtsi, Ukraine

Irina V. Soltys
Yuriy Fedkovych Chernivtsi National University
Chernivtsi, Ukraine

Benjamin Hogan
Opto-Electronics and Measurement Techniques Laboratory
University of Oulu
Oulu, Finland

# Introduction

The objective of optical laser diagnostics of the properties of biological tissues is to establish the photometric, polarization, and correlation structure of the coherent fields formed as the light passes through the tissue [1–10]. The structure of such fields can be estimated on the basis of various correlation approximations. For scalar fields, a "two-point" correlation parameter known as the "degree of coherence" is introduced. The degree of coherence is a measure of the correlation between the parallel amplitude components at two different points [11–70]. For polarization fields, a further ("single-point") correlation parameter is introduced—the degree of polarization. The degree of polarization is described by the coherence matrix [12–16, 55, 56, 75, 76] and characterizes the degree of correlation between the orthogonal components of the amplitude of electric field at a single point of the laser field.

These separate parametric descriptions have been generalized by the correlation approach to the analysis of vector fields of coherent radiation. In doing so, new correlation ("complex degree of coherence"—CDC) [66, 67] and polarization-correlation ("complex degree of mutual polarization"—CDMP) parameters [69–72] are used. Subsequently, the indicated correlation approaches were combined in the framework of laser polarimetry of microscopic images of histological sections of biological tissues by introducing a new parameter—the "complex degree of mutual anisotropy"—CDMA [73]. The possibility of using the CDMA in the analysis of optical anisotropy of birefringent networks of biological tissues has been demonstrated in [74–83].

A significant breakthrough in the development of correlation methods for the optical diagnosis of biological tissues was the successful combination of laser polarimetry and autofluorescence methods [12–14]. It holds great promise for laser-polarimetric and autofluorescence diagnostics of the optical anisotropy of biological tissues. This requires the development of new methods, based on the use of a universal polarization-correlation approach, in order to:

- Conduct scale-selective wavelet and Fourier analyses [84–86] of polarization-inhomogeneous fields of the main types of biological tissue [11, 14–16, 22–26, 35, 44, 49–54]. This can be considered as a "single-point" laser-polarimetric correlation approach.

- Perform comparative studies of the coordinate distributions of the values of the CDMP [71–74] and CDMA [75–83] parameters of biological tissues and their images. In this case, we use a "two-point" polarization-correlation approach.
- Develop wavelet methods and Fourier analysis [86] of coordinate distributions of "two-point" polarization-correlation parameters CDMP and CDMA. This constitutes an integrated polarization-correlation approach.

The solutions to such problems will provide the foundations of a new universal method of Stokes correlometry, combining the principles of polarization [14–16, 22–26], polarization-correlation [71–83], and autofluorescence correlation [12–14] analysis of the interaction of laser radiation with optically anisotropic biological tissues.

Therefore, the studies described in this book are both relevant and important to the need to extend the arsenal of diagnostic methods [17–20, 27–29, 31, 36, 38, 39, 43–50, 57–59]. The book presents advances in the studies of biological objects by using a Stokes-correlometric approach for the analysis of optical anisotropy of biological tissues. Complex wavelet, Fourier, and polarization-correlation analyses of the coordinate distributions of both the azimuth and the polarization ellipticity of polarization-inhomogeneous images of biological layers are considered, as well as the elements of the Mueller matrix of polycrystalline networks. This enables the development of complex optical-physical methods for the diagnosis of birefringence changes due to necrotic conditions of human organs.

The overarching purpose of this book is to describe new informative diagnostic parameters for characterizing polycrystalline networks of biological layers and the laser fields scattered by them. This is achieved by developing methods for azimuthally invariant laser Stokes correlometry using scale-selective wavelet and Fourier analyses of the topographic structure of polarization, Mueller matrix, and polarization-correlation images of biological tissues.

The diagnostic capabilities of the wavelet analysis of two-dimensional polarization maps and Mueller-matrix invariants characterizing the birefringence of myocardial fibrillar networks, deceased due to mechanical asphyxiation and heart attack, are investigated. The sensitivity of the third- and fourth-order statistical moments characterizing the distributions of the amplitudes of the wavelet-coefficients, at different scales of the MHAT function, to change in the distribution of phase shifts of linearly and circularly birefringent myosin fibers is established. The method of spatial-frequency filtering of microscopic images of histological sections of biological tissues, in the Fourier plane of a polarization-inhomogeneous object field, and under conditions of azimuthally invariant polarization and Mueller-matrix mapping, was tested for the first time. The relationships between the values of the statistical moments of the third and fourth orders, characterizing the distributions of polarization parameters and elements of the Mueller matrix of histological myocardium sections, and changes in the birefringence of polycrystalline networks caused by necrotic changes are considered. Based on the polarization-correlation approach to the description of object laser fields formed by spatially ordered and disordered optically anisotropic networks of biological crystals, a comparative study of the diagnostic effectiveness of statistical analysis methods for CDMP mapping of the azimuth distributions and elliptic polarization, as well as CDMP mapping of the distributions of directions of the optical axes and phase shifts introduced by birefringent fibrils, is carried out for the first time. A novel, scale-selective, wavelet analysis of the spatial distributions of the values of the CDMP and CDMA modulus of microscopic images of optically anisotropic histological sections of biological tissues with structured birefringent fibrillar networks was proposed. The method of scale-selective polarization-correlation mapping using spatial-frequency filtering of the spatial distributions of the values of the CDMP and CDMA modulus of microscopic images of optically anisotropic histological sections of biological tissues with structured birefringent fibrillar networks is developed.

## References


1. V. Tuchin, L. Wang, D. Zimnjakov, *Optical Polarization in Biomedical Applications* (Springer, New York, USA, 2006)
2. R. Chipman, *Polarimetry*. in ed. by M. Bass. Handbook of Optics: Vol I—Geometrical and Physical Optics, Polarized Light, Components and Instruments (McGraw-Hill Professional, New York, 2010), pp. 22.1–22.37
3. N. Ghosh, M. Wood, A. Vitkin, in *Polarized Light Assessment of Complex Turbid Media such as Biological Tissues Via Mueller Matrix Decomposition*, ed. by V. Tuchin. Handbook of Photonics for Biomedical Science (CRC Press, Taylor & Francis Group, London, 2010), pp. 253–282
4. S. Jacques, Polarized Light Imaging of Biological Tissues, in *Handbook of Biomedical Optics*. ed. by D. Boas, C. Pitris, N. Ramanujam (CRC Press, Boca Raton, London, New York, 2011), pp. 649–669
5. N. Ghosh, Tissue polarimetry: concepts, challenges, applications, and outlook. J. Biomed. Opt. **16**(11), 110801 (2011)
6. M. Swami, H. Patel, P. Gupta, Conversion of 3x3 Mueller matrix to 4x4 Mueller matrix for non-depolarizing samples. Opt. Commun. **286**, 18–22 (2013)
7. D. Layden, N. Ghosh, A. Vitkin, in *Quantitative Polarimetry for Tissue Characterization and Diagnosis*, ed. by R. Wang, V. Tuchin. Advanced Biophotonics: Tissue Optical Sectioning (CRC Press, Taylor & Francis Group, Boca Raton, London, New York, 2013), pp. 73–108
8. T. Vo-Dinh, *Biomedical Photonics Handbook*, 3 vol. set, 2nd edn. (CRC Press, Boca Raton, 2014)
9. A. Vitkin, N. Ghosh, A. De Martino, Tissue Polarimetry, in *Photonics: Scientific Foundations, Technology and Applications*, 4th edn., ed. by D. Andrews (Wiley, Hoboken, New Jersey, 2015), pp. 239–321
10. V. Tuchin, *Tissue Optics: Light Scattering Methods and Instruments for Medical Diagnosis*, 2nd edn. (SPIE Press, Bellingham, Washington, USA, 2007).
11. P. Li, H. R. Lee, S. Chandel, C. Lotz, F. Kai Groeber-Becker, S. Dembski, R. Ossikovski, H. Ma, and T. Novikova Analysis of tissue microstructure with Mueller microscopy: logarithmic decomposition and Monte Carlo modeling, J. Biomed. Opt. **25**(1), 015002 (2020)
12. H. R. Lee, P. Li, T. S. H. Yoo, C. Lotz, F. Kai Groeber-Becker, S. Dembski, E. Garcia-Caurel, R. Ossikovski, H. Ma, and T. Novikova Digital histology with Mueller microscopy: how to mitigate an impact of tissue cut thickness fluctuations, J. Biomed. Opt. **24**(7) 076004 (2019)
13. H. R. Lee, C. Lotz, F. Kai Groeber-Becker, S. Dembski, E. Garcia-Caurel, R. Ossikovski, T. Novikova Mueller microscopy of full thickness skin models combined with image segmentation, Proc. SPIE Advances in Microscopic Imaging II 11076, 1107615 (2019)

14. H. R. Lee, T. S. H. Yoo, P. Li, C. Lotz, F. Kai Groeber-Becker, S. Dembski, E. Garcia-Caurel, R. Ossikovski, T. Novikova Mueller microscopy of anisotropic scattering media: theory and experiments, Proc. SPIE 10677, Unconventional Optical Imaging, 1067718 (2018)
15. T. Novikova, J. Rehbinder, H. Haddad, S. Deby, B. Teig, A. Nazac, A. Pierangelo, F. Moreau, A. De Martino, Multispectral Mueller Matrix Imaging Polarimetry for Studies of Human Tissue, OSA Biophotonics Congress, Clinical and Translational Biophotonics, paper TTh3B (2016)
16. W. Bickel, W. Bailey, Stokes vectors, Mueller matrices, and polarized scattered light. Am. J. Phys. **53**(5), 468–478 (1985)
17. A. Doronin, C. Macdonald, I. Meglinski, Propagation of coherent polarized light in turbid highly scattering medium. J. Biomed. Opt. **19**(2), 025005 (2014)
18. A. Doronin, A. Radosevich, V. Backman, I. Meglinski, Two electric field Monte Carlo models of coherent backscattering of polarized light. J. Opt. Soc. America A **31**(11), 2394 (2014)
19. A. Ushenko, V. Pishak, in *Laser Polarimetry of Biological Tissue: Principles and Applications*, ed. by V. Tuchin. Handbook of Coherent-Domain Optical Methods: Biomedical Diagnostics (Environmental and Material Science, 2004), pp. 93–138
20. O. Angelsky, A. Ushenko, Y. Ushenko, V. Pishak, A. Peresunko, in *Statistical, Correlation and Topological Approaches in Diagnostics of the Structure and Physiological State of Birefringent Biological Tissues*. Handbook of Photonics for Biomedical Science (2010), pp. 283–322
21. Y. Ushenko, T. Boychuk, V. Bachynsky, O. Mincer, in *Diagnostics of Structure and Physiological State of Birefringent Biological Tissues: Statistical, Correlation and Topological Approaches*, ed. by V. Tuchin. Handbook of Coherent-Domain Optical Methods (Springer Science+Business Media, 2013)
22. O. Angelsky, A. Ushenko, Y. Ushenko, Investigation of the correlation structure of biological tissue polarization images during the diagnostics of their oncological changes. Phys. Med. Biol. **50**(20), 4811–4822 (2005)
23. V. Ushenko, O. Dubolazov, A. Karachevtsev, Two wavelength Mueller matrix reconstruction of blood plasma films polycrystalline structure in diagnostics of breast cancer. Appl. Opt. **53** (10), B128 (2016)
24. Y. Ushenko, G. Koval, A. Ushenko, O. Dubolazov, V. Ushenko, O. Novakovskaia, Mueller-matrix of laser-induced autoffiuorescence of polycrystalline films of dried peritoneal fluid in diagnostics of endometriosis. J. Biomed. Opt. **21**(7), 071116 (2016)
25. A. Ushenko, A. Dubolazov, V. Ushenko, O. Novakovskaya, Statistical analysis of polarization-inhomogeneous Fourier spectra of laser radiation scattered by human skin in the tasks of differentiation of benign and malignant formations. J. Biomed. Opt. **21**(7), 071110 (2016)
26. V. Ushenko, N. Pavlyukovich, L. Trifonyuk, Spatial-frequency azimuthally stable cartography of biological polycrystalline networks. Int. J. Opt. **2013**, 1–7 (2013)
27. S. Manhas, M.K. Swami, P. Buddhiwant, N. Ghosh, P.K. Gupta, K. Singh, Mueller matrix approach for determination of optical rotation in chiral turbid media in backscattering geometry. Opt. Exp. **14**, 190–202 (2006)
28. Y. Deng, S. Zeng, Q. Lu, Q. Luo, Characterization of backscattering Mueller matrix patterns of highly scattering media with triple scattering assumption. Opt. Exp. **15**, 9672–9680 (2007)
29. S.Y. Lu, R.A. Chipman, Interpretation of Mueller matrices based on polar decomposition. J. Opt. Soc. Am. A **13**, 1106–1113 (1996)
30. Y. Guo, N. Zeng, H. He, T. Yun, E. Du, R. Liao, H. Ma, A study on forward scattering Mueller matrix decomposition in anisotropic medium. Opt. Exp. **21**, 18361–18370 (2013)
31. A. Pierangelo, S. Manhas, A. Benali, C. Fallet, J.L. Totobenazara, M.R. Antonelli, T. Novikova, B. Gayet, A. De Martino, P. Validire, Multispectral Mueller polarimetric imaging detecting residual cancer and cancer regression after neoadjuvant treatment for colorectal carcinomas. J. Biomed. Opt. **18**, 046014 (2013)
32. V.P. Ungurian, O.I. Ivashchuk, V.O. Ushenko, Statistical analysis of polarizing maps of blood plasma laser images for the diagnostics of malignant formations. Proc. SPIE **8338**, 83381L (2011)

33. V.A. Ushenko, O.V. Dubolazov, A.O. Karachevtsev, Two wavelength Mueller matrix reconstruction of blood plasma films polycrystalline structure in diagnostics of breast cancer. Appl. Opt. **53**, B128–B139 (2014)
34. V.P. Prysyazhnyuk, Y.A. Ushenko, A.V. Dubolazov, A.G. Ushenko, V.A. Ushenko, Polarization-dependent laser autofluorescence of the polycrystalline networks of blood plasma films in the task of liver pathology differentiation. Appl. Opt. **55**, B126–B132 (2016)
35. V.A. Ushenko, M.S. Gavrylyak, Azimuthally invariant Mueller-matrix mapping of biological tissue in differential diagnosis of mechanisms protein molecules networks anisotropy. Proc. SPIE **8812**, 88120Y (2013)
36. V.A. Ushenko, M.P. Gorsky, Complex degree of mutual anisotropy of linear birefringence and optical activity of biological tissues in diagnostics of prostate cancer. Opt. Spectrosc. **115**, 290–297 (2013)
37. V.A. Ushenko, A.V. Dubolazov, Correlation and self similarity structure of polycrystalline network biological layers Mueller matrices images. Proc. SPIE **8856**, 88562D (2013)
38. V.O. Ushenko, Spatial-frequency polarization phasometry of biological polycrystalline networks. Opt. Mem. Neur. Netw. **22**, 56–64 (2013)
39. V.A. Ushenko, N.D. Pavlyukovich, L. Trifonyuk, Spatial-frequency azimuthally stable cartography of biological polycrystalline networks. Int. J. Opt. **683174**, 2013 (2013)
40. V. Devlaminck, Physical model of differential Mueller matrix for depolarizing uniform media. J. Opt. Soc. America A **30**(11), 2196 (2013)
41. Y.A. Ushenko, Spatial-frequency Fourier polarimetry of the complex degree of mutual anisotropy of linear and circular birefringence in the diagnostics of oncological changes in morphological structure of biological tissues. Quantum. Electron. **42**, 727–732 (2012)
42. V.A. Ushenko, Complex degree of mutual coherence of biological liquids. Proc. SPIE **8882**, 88820V (2013)
43. V.A. Ushenko, M.P. Gorsky, Complex degree of mutual anisotropy of linear birefringence and optical activity of biological tissues in diagnostics of prostate cancer. Opt. Spectrosc. **115**, 290–297 (2013)
44. Y.A. Ushenko, Jones-matrix mapping of complex degree of mutual anisotropy of birefringent protein networks during the differentiation of myocardium necrotic changes. Appl. Opt. **55**, B113-B119 (2016)
45. L. Cassidy, Basic concepts of statistical analysis for surgical research. J. Surg. Res. **128**(2), 199–206 (2005)
46. C.S. Davis, *Statistical Methods of the Analysis of Repeated Measurements* (Springer, New York, 2002).
47. A. Petrie, C. Sabin, *Medical Statistics at a Glance* (Wiley-Blackwell, Chichester, UK, 2009).
48. A.G. Ushenko, A. Dubolazov, V. Ushenko, O. Novakovskaya, Statistical analysis of polarization-inhomogeneous Fourier spectra of laser radiation scattered by human skin in the tasks of differentiation of benign and malignant formations. J. Biomed. Opt. **21**(7), 071110 (2016)
49. Y.A. Ushenko, G.D. Koval, A.G. Ushenko, O.V. Dubolazov, V.A. Ushenko, O.Y. Novakovskaia, Mueller-matrix of laser-induced autofluorescence of polycrystalline films of dried peritoneal fluid in diagnostics of endometriosis. J. Biomed. Opt. **21** (7), 071116 (2016)
50. A.G. Ushenko, O.V. Dubolazov, V.A. Ushenko, O.Y. Novakovskaya, O.V. Olar, Fourier polarimetry of human skin in the tasks of differentiation of benign and malignant formations. Appl. Opt. **55**(12), B56–B60 (2016)
51. Y.A. Ushenko, V.T. Bachynsky, O.Y. Vanchulyak, A.V. Dubolazov, M.S. Garazdyuk, V.A. Ushenko, Jones-matrix mapping of complex degree of mutual anisotropy of birefringent protein networks during the differentiation of myocardium necrotic changes. Appl. Opt. **55** (12), B113–B119 (2016)
52. A.V. Dubolazov, N.V. Pashkovskaya, Y.A. Ushenko, Y.F. Marchuk, V.A. Ushenko, O.Y. Novakovskaya, Birefringence images of polycrystalline films of human urine in early diagnostics of kidney pathology. Appl. Opt. **55**(12), B85–B90 (2016)

53. M.S. Garazdyuk, V.T. Bachinskyi, O.Y. Vanchulyak, A.G. Ushenko, O.V. Dubolazov, M.P. Gorsky, Polarization-phase images of liquor polycrystalline films in determining time of death. Appl. Opt. **55**(12), B67–B71 (2016)
54. M. Borovkova, M. Peyvasteh, O. Dubolazov, Y. Ushenko, V. Ushenko, A. Bykov, S. Deby, J. Rehbinder, T. Novikova, I. Meglinski, Complementary analysis of Mueller-matrix images of optically anisotropic highly scattering biological tissues. J. Eur. Opt. Soc. **14**(1), 20 (2018)
55. V. Ushenko, A. Sdobnov, A. Syvokorovskaya, A. Dubolazov, O. Vanchulyak, A. Ushenko, Y. Ushenko, M. Gorsky, M. Sidor, A. Bykov, I. Meglinski, 3D Mueller-matrix diffusive tomography of polycrystalline blood films for cancer diagnosis. Photonics **5**(4), 54 (2018)
56. L. Trifonyuk, W. Baranowski, V. Ushenko, O. Olar, A. Dubolazov, Y. Ushenko, B. Bodnar, O. Vanchulyak, L. Kushnerik, M. Sakhnovskiy, 2D-Mueller-matrix tomography of optically anisotropic polycrystalline networks of biological tissues histological sections. Opto-Electron. Rev. **26**(3), 252–259 (2018)
57. V.A. Ushenko, A.V. Dubolazov, L.Y. Pidkamin, M.Y. Sakchnovsky, A.B. Bodnar, Y.A. Ushenko, A.G. Ushenko, A. Bykov, I. Meglinski, Mapping of polycrystalline films of biological fluids utilizing the Jones-matrix formalism. Laser Phys. **28**(2), 025602 (2018)
58. V.A. Ushenko, A.Y. Sdobnov, W.D. Mishalov, A.V. Dubolazov, O.V. Olar, V.T. Bachinskyi, A.G. Ushenko, Y.A. Ushenko, O.Y. Wanchuliak, I. Meglinski, Biomedical applications of Jones-matrix tomography to polycrystalline films of biological fluids. J. Innovative Opt. Health Sci. **12**(6), 1950017 (2019)
59. M. Borovkova, L. Trifonyuk, V. Ushenko, O. Dubolazov, O. Vanchulyak, G. Bodnar, Y. Ushenko, O. Olar, O. Ushenko, M. Sakhnovskiy, A. Bykov, I. Meglinski, Mueller-matrix-based polarization imaging and quantitative assessment of optically anisotropic polycrystalline networks. PLoS ONE **14**(5), e0214494 (2019)
60. A. Ushenko, A. Sdobnov, A. Dubolazov, M. Grytsiuk, Y. Ushenko, A. Bykov, I. Meglinski, Stokes-correlometry analysis of biological tissues with polycrystalline structure. IEEE J. Sel. Top. Quantum Electron. **25**(1), 8438957 (2019)
61. O. Vanchulyak, O. Ushenko, V. Zhytaryuk, V. Dvorjak, O. Pavlyukovich, O. Dubolazov, N. Pavlyukovich, N.P. Penteleichuk, Stokes-correlometry of polycrystalline films of biological ffiuids in the early diagnostics of system pathologies. Proc. SPIE—Int. Soc. Opt. Eng. **11105**, 1110519 (2019)
62. A.V. Dubolazov, O.V. Olar, L.Y. Pidkamin, A.D. Arkhelyuk, A.V. Motrich, V.T. Bachinskiy, O.V. Pavliukovich, N. Pavliukovich, Differential components of Muller matrix partially depolarizing biological tissues in the diagnosis of pathological and necrotic changes. Proc. SPIE **11087**, 1108713 (2019)
63. O. Ushenko, V. Zhytaryuk, V. Dvorjak, I.V. Martsenyak, O. Dubolazov, B.G. Bodnar, O.Y. Vanchulyak, S. Foglinskiy, Multifunctional polarization mapping system of networks of biological crystals in the diagnostics of pathological and necrotic changes of human organs. Proc. SPIE **11087**, 110870S (2019)
64. O. Pavlyukovich, N. Pavlyukovich, Y. Ushenko, O. Galochkin, M. Sakhnovskiy, M. Kovalchuk, A. Dovgun, S. Golub, O. Dubolazov, Fractal analysis of patterns for birefringence biological tissues in the diagnostics of pathological and necrotic states. Proc. SPIE **11105**, 1110518 (2019)
65. A.V. Dubolazov, O.V. Olar, L.Y. Pidkamin, A.D. Arkhelyuk, A.V. Motrich, M.V. Shaplavskiy, B.G. Bodnar, Y. Sarkisova, N. Penteleichuk, Polarization-phase reconstruction of polycrystalline structure of biological tissues. Proc. SPIE. **11087**, 1108714 (2019)
66. A.V. Dubolazov, O.V. Olar, L.Y. Pidkamin, A.D. Arkhelyuk, A.V. Motrich, O. Petrochak, V.T. Bachynskiy, O. Litvinenko, S. Foglinskiy, Methods and systems of diffuse tomography of optical anisotropy of biological layers. Proc. SPIE. **11087**, 110870P (2019)
67. E. Wolf, Unified theory of coherence and polarization of random electromagnetic beams. Phys. Lett. A **312**, 263–267 (2003)
68. J. Tervo, T. Setala, A. Friberg, Degree of coherence for electromagnetic. Opt. Express **11**, 1137–1143 (2003)

69. J.M. Movilla, G. Piquero, R. Martínez-Herrero, P.M. Mejías, Parametric characterization of non-uniformly polarized. Opt. Commun. **149**, 230–234 (1998)
70. J. Ellis, A. Dogariu, Complex degree of mutual polarization. Opt. Lett. **29**, 536–538 (2004)
71. C. Mujat, A. Dogariu, Statistics of partially coherent beams: a numerical analysis. J. Opt. Soc. Am. A **21**(6), 1000–1003 (2004)
72. F. Gori, Matrix treatment for partially polarized, partially coherent beams. Opt. Lett. **23**, 241–243 (1998)
73. E. Wolf, Significance and measurability of the phase of a spatially coherent optical field. Opt. Lett. **28**, 5–6 (2003)
74. M. Mujat, A. Dogariu, Polarimetric and spectral changes in random electromagnetic fields. Opt. Lett. **28**, 2153–2155 (2003)
75. J. Ellis, A. Dogariu, S. Ponomarenko, E. Wolf, Interferometric measurement of the degree of polarization and control of the contrast of intensity ffiuctuations. Opt. Lett. **29**, 1536–1538 (2004)
76. O. Angelsky, A. Ushenko, Y. Ushenko, Complex degree of mutual polarization of biological tissue coherent images for the diagnostics of their physiological state. J. Biomed. Opt. **10**(6), 060502 (2005)
77. Y. Ushenko, Complex degree of mutual polarization of Biotissue's Speckle-images. Ukr. J. Phys Opt. **6**(3), 104–113 (2005)
78. S.B. Yermolenko, C.Y. Zenkova, A.-P. Angelskiy, Polarization manifestations of correlation (intrinsic coherence) of optical fields. Appl. Opt. **47**(32) (2008)
79. O.V. Angelsky, A.G. Ushenko, A.O. Angelskaya, Y.A. Ushenko, Correlation- and singular-optical approaches in diagnostics of polarization inhomogeneity of coherent optical fields from biological tissues. Ukr. J. Phys. Opt. **8**(2), 106–123 (2007)
80. O.V. Angelsky, A.G. Ushenko, A.O. Angelskaya, Y.A. Ushenko, Polarization correlometry of polarization singularities of biological tissues object fields. Proc. SPIE **6616**, 1–9 (2007)
81. Y.O. Ushenko, Y.Y. Tomka, O.I. Telenga, I.Z. Misevitch, V.V. Istratiy, Complex degree of mutual anisotropy of biological liquid crystals nets. Opt. Eng. **50**, 039001 (2011)
82. Y.A. Ushenko, O.I. Telenga, A.P. Peresunko, O.K. Numan, New parameter for describing and analyzing the optical-anisotropic properties of biological tissues. J. Innov. Opt. Health Sci. **4**(4), 463–475 (2011)
83. A.V. Dubolazov, O.Y. Telenha, V.A. Ushenko, M.I. Sydor, Characteristic values of Mueller-matrixes images of biological liquid crystals net for diagnostics of human tissues anisotropy. Proc. SPIE **8338**, 83380Z (2011)
84. O.Y. Novakovska, Polarization correlometry of characteristic states of Muller-matrix images of phase-inhomogeneous biological layers. Semicond. Phys. Quantum Electron. Optoelectron. **15**(3), 230–237 (2012)
85. A.G. Ushenko, Y.A. Ushenko, Y.Y. Tomka, O.V. Dubolazov, O.Y. Telenga, V.I. Istratiy, A.O. Karachevtsev, The interconnection between the coordinate distribution of Muller-matrixes images characteristics values of biological liquid crystals net and the pathological changes of human tissues : 12–16 July 2010, 9th Euro-American Workshop on Information Optics. Helsinki, Finland (2010)
86. Y.A. Ushenko, O.V. Dubolazov, O.Y. Telenga, A.P. Angelsky, A.O. Karachevtcev, V.Balanetska, Complex degree of mutual anisotropy of biological liquid crystals net. Proc. SPIE **8087**, 80872Q (2011)
87. O.V. Angelsky, Y.A. Ushenko, V.O. Balanetska, The degree of mutual anisotropy of biological liquids polycrystalline nets as a parameter in diagnostics and differentiations of hominal inffiammatory processes. Proc. SPIE **8338**, 83380S (2011)
88. Y.A. Ushenko, A.V. Dubolazov, A.O. Karachevtcev, N.I. Zabolotna, Complex degree of mutual anisotropy in diagnostics of biological tissues physiological changes. Proc. SPIE **8134**, 81340O (2011)
89. Y.Y. Tomka, Wavelet analysis of biological tissue's Mueller-matrix images. Proc. SPIE **7008**, 700823 (2008)

90. O.V. Dubolazov, Y.O. Ushenko, Y.Y. Tomka, O.G. Pridiy, A.V. Motrich, I.Z. Misevitch, V.V. Istratiy, Wavelet analysis for Mueller matrix images of biological crystal networks. Semicond. Phys. Quantum Electron. Optoelectron. **12**(4), 391–398 (2009)
91. A.O. Karachevtsev, Fourier Stokes-polarimetry of biological layers polycrystalline networks. Semicond. Phys. Quantum Electron. Optoelectron. **15**(3), 252–268 (2012)

# Contents

# Chapter 1
# Methods and Means of Polarization Correlation of Fields of Laser Radiation Scattered by Biological Tissues

## 1.1 Polarization-Inhomogeneous Fields and Methods for Their Analysis

Any physical object or medium is a complex, optically heterogeneous, structure. Such heterogeneity is characterized by the spatial distributions of the overall optical parameters (refractive indices and absorption) and their anisotropic components (linear and circular birefringence, linear and circular dichroism). In addition, there can be spatial and/or angular heterogeneity of the orientation, size, shape, and volume of the particles or domains constituting the overall physical body at the micro- and macrolevels [1–6]. A common feature of such optically heterogeneous objects is scattering of incident light. This scattering is the subject of optics of scattering media.

Light interacting with biological objects is therefore scattered by biological structures. These can include various biological fluids (blood, lymph, cerebrospinal fluid, urine, bile, etc.) and tissues (muscle, connective, epithelial, nervous). All these biological objects typically exhibit multiple scattering and form optically turbid layers [7, 8]. The optical properties of such objects within the photometric approximation are described by the model of multiple scattering of scalar waves in a randomly inhomogeneous medium with absorption [1, 5–9]. If the optical thickness of a biological layer is insignificant (weakly or singly scattering media), then its properties with respect to light intensity conversion are traditionally determined with a model of single scattering within an ordered medium with densely packed scattering centers [10]. From the point of view of the geometric construction of such optically inhomogeneous layers, one can distinguish surface or rough scattering [10–12], as well as volume scattering [13–15].

Historically, there are three main approaches to the studies of the optical properties of such biological objects—1) spectrophotometric, that is based on the analysis of spatial or temporal changes in the intensity of scattered radiation in the optical range of electromagnetic wavelengths [16–19]; 2) polarimetric, based on the analysis of the distributions of the azimuth and polarization elliptic distributions, or the degree



I. Meglinski et al., *Shedding the Polarized Light on Biological Tissues*,
SpringerBriefs in Applied Sciences and Technology,
https://doi.org/10.1007/978-981-10-4047-4_1

of polarization of the scattered radiation field [1, 6, 20, 21]; 3) correlation, which determines the degree of correlation between parallel components of the amplitude of polarized electromagnetic waves at various points of the scattered field of optical radiation [22–30].

Applying such methods we obtain a combination of the intensity distributions, polarization parameters (both azimuth and ellipticity angles of polarization ellipse), and phase of the scattered radiation field within the optical wavelength range.

The obtained distributions of optical characteristics that are generated by biological layers are traditionally analyzed using a statistical approach. For example, the scattering on rough surface is described by a set of statistical moments characterizing the distribution of the heights of surface microroughness. The statistical amplitude-phase moments of the scattered radiation field are experimentally measured [14].

For volume scattering objects (biological tissues), the process of propagation of optical radiation is accompanied by a change in all its spectrophotometric, polarization and correlation characteristics. The main factors that induce these changes are reflection, refraction of radiation at the media interfaces, diffraction by local microinhomogeneities, interference of laser waves scattered by such microstructures, linear and circular birefringence, and absorption dichroism [11, 31–43].

The complexity of light scattering process requires the development of various methods of optical diagnostics of objects from both practical and fundamental point of view. In order to overcome the ambiguity of the analysis of the obtained data, the most important is to get the complete information on the spectrophotometric polarization or correlation distributions of the optical parameters. In other words, further progress in biomedical diagnostics is associated with the development of integrated "interdisciplinary" methods. These primarily include new approaches to the study of scattered radiation fields, for example, fractal optics [12, 22, 23], or singular optics [13, 14, 44–48]. The development of these research fields built the foundation of a new branch of biomedical optics, namely, laser polarimetry of optically anisotropic biological tissues [16–21, 31–43, 49–72]. The use of lasers as the light sources ensured a new development of the nephelometry method that is widely used in biology and medicine, by significantly simplifying the measurements and increasing their reliability in immunology, virology, and hematology [12–14, 18, 19, 31–35, 49–56].

If we draw a further parallel, we can state that both polarization nephelometry and laser polarimetry share a common approach—the analysis of the scattering matrix of a biological object or its Mueller matrix. It is known that this matrix of a linear operator characterizes the microstructure of various biological objects (the geometric and statistical parameters of individual scattering particles, the presence of phase and amplitude anisotropy) [38–43]. The information obtained on the microstructure of biological tissues enables, for example, early diagnosis of cataracts and glaucoma [1].

The main feature of the combination of various methods of laser polarimetry (polarization, phase, Mueller-matrix mapping) is the analysis of the obtained data using the optical-geometric model of biological tissues of various types that mimics “frozen” optically—uniaxial liquid crystals.

Such a model approach was developed in [16, 17, 20, 21], where a layer of biological tissue is considered as a two-component amorphous-polycrystalline structure. The amorphous components of biological tissue (fats, lipids, unstructured proteins) are optically isotropic. The polycrystalline components of biological tissue are presented by fibrillar proteins (such as collagen proteins, elastin, fibrin, myosin) network [13]. The optical properties of each individual fibril are described by the matrix operator of an optically uniaxial birefringent crystal [11–14].

$$\{Q\} = \left\| \begin{matrix} 1; & 0; & 0; & 0; \\ 0; & \left(\sin^2\frac{\delta}{2}\cdot\cos 2\rho + \cos^2\frac{\delta}{2}\right); & \left(0,5\sin 4\rho\sin^2\frac{\delta}{2}\right); & (-\sin 2\rho\sin\delta); \\ 0; & \left(0,5\sin 4\rho\sin^2\frac{\delta}{2}\right); & \left(-\sin^2\frac{\delta}{2}\cdot\cos 2\rho + \cos^2\frac{\delta}{2}\right); & (\cos 2\rho\sin\delta); \\ 0; & (\sin 2\rho\sin\delta); & (-\cos 2\rho\sin\delta); & \left(2\cos^2\frac{\delta}{2} - 1\right); \end{matrix} \right\|. \tag{1.1}$$

Here, $\rho$ is the orientation of the optical axis determined by the direction of the long axis of the fibrils, and $\delta$ is the magnitude of the phase shift between the orthogonal components of the amplitude of electric field of elecromagnetic wave.

The use of this model-based analysis of the phenomenon of light scattering by optically thin biological tissues made it possible to explain the mechanisms of formation of polarization-inhomogeneous object fields of biological tissues of various types, in particular histological sections of bone and muscle tissue, as well as multilayer tissue of a woman’s reproductive system (e.g. myometrium).

On this basis, algorithms are defined for describing the relationships between the polarization state (azimuth and ellipticity angles) of detected light at any point of a microscopic image and the corresponding fibrillar direction (optical axis of the crystal) and the linear birefringence of the biological sample [32–35]. This result made it possible to create a method for visualizing birefringent networks of biological tissues. In addition, a statistical analysis has been introduced to characterize the obtained images—the calculation of a set of statistical moments of the first to fourth orders characterizes the spatial distribution of polarization images of fibrillar networks of histological sections of biological tissue [18, 19, 32–35, 49, 50].

A new step in the development of laser polarimetry was experimental mapping of the distribution of polarization states in microscopic images of biological samples.

## 1.2 Polarization Mapping of Microscopic Images of Biological Tissues

In [20, 21], an optical scheme for polarization mapping is shown and described (Fig. 1.1).

The biological tissue sample was illuminated with a parallel ($\varnothing = 10^4$ μm) He–Ne laser beam ($\lambda = 0.6328$ μm, $W = 5.0$ mW) generated by a source 1 and passed through a collimator 2. Explanations on element 1 (source) and beam expander lenses 2. The modulation of the polarization state of the incident beam was performed with quarter-wave plates 3 & 5 and polarizer 4. The microlens 7 projected the image of the histological section 6 into the plane of the photosensitive area (800 × 600 pixels) of the CCD camera 10, passing through a polarizer 9 and a quarter-wave plate 8 on the way.

In [56, 62] the relationship was found between the statistical structure of polarization maps of microscopic images of histological sections of biological tissues of various types and the features of the morphological structure of their birefringence. It was found that for CT, the main birefringent structures are large-scale domains of spatially ordered collagen fibers. The optical manifestation of such a structure of the biological layer is revealed in the formation of a uniform polarization signature within such domains.

The studies [18, 19, 37–39, 49, 50, 56, 60–62, 66–69] provided the criteria for the polarization differentiation of optical anisotropy of histological sections of physiologically normal and pathologically altered human organ tissues.

The successful application of the method of polarization mapping stimulated the development of more complex methods aimed at increasing the sensitivity of laser polarimetry to changes in the optical anisotropy of biological layers. One of the main directions of such improvement was the use of scale-selective wavelet analysis of two-dimensional distributions of polarization states in microscopic images of biological tissue samples.

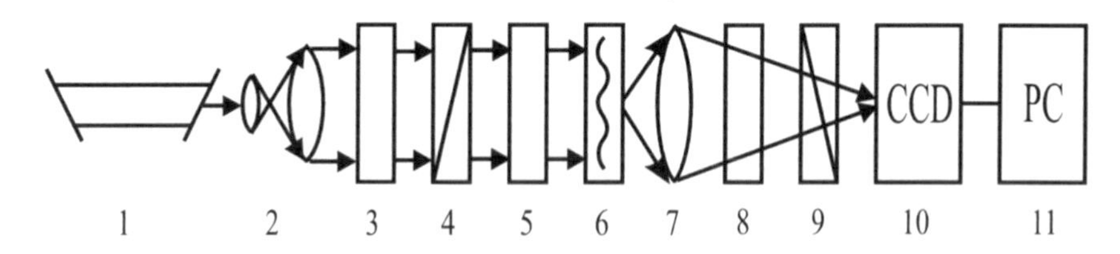


**Fig. 1.1** Optical measurement scheme. Explanation in the text [11–13, 20, 21]

## 1.3 Wavelet Analysis of Polarization Maps and Mueller-Matrix Images of Biological Tissues

The use of a mathematical technique—wavelet analysis—allows one to study the spatial distribution of the azimuth and ellipticity angles of polarization ellipse. This analysis can be done for the microscopic images of polycrystalline networks of the biological layers, corresponding to a predetermined range of changes in the geometric dimensions of the crystals. This technique was applied in [12] to study the properties of polycrystalline films of human blood plasma.

In [72–74], information is provided on the algorithm for applying the wavelet analysis of spatial distributions of polarization parameters in microscopic images of polycrystalline films of dried blood plasma.

A two-dimensional set of data $\begin{pmatrix} 11 & .. & 1m \\ . & .. & . \\ . & .. & . \\ 1n & .. & mn \end{pmatrix}$ recorded using a CCD camera, with pixels indexed in the x-direction from 0 to $n$ with a step $b$, is treated with the help of the wavelet functions whose window dimensions $a$ vary in the required range.

The result of this operation is a set of calculated wavelet coefficients

$$W_{a,b} = \begin{pmatrix} W(a_{\min}, b_1) & \cdot\cdot & W(a_{\min}, b_m) \\ . & .. & . \\ . & .. & . \\ W(a_{\max}, b_1) & \cdot\cdot & W(a_{\max}, b_m) \end{pmatrix} \tag{1.2}$$

for each line $(k1;\ km)$ of the photosensitive area of the CCD camera.

This operation is sequentially repeated for the full set of lines from 0 to $m$ of the sensor area of the CCD camera. The resulting set of wavelet coefficients $W_{a,b}(m \times n)$ is then averaged [73, 74]

$$\overline{W}_{a,b} = \begin{pmatrix} \overline{W}(a_{\min}, b_1) = \frac{\sum_{j=1}^{m} W_j(a_{\min}, b_1)}{m}; & .. & \overline{W}(a_{\min} b = m) = \frac{\sum_{j=1}^{m} W_j(a_{\min}, b_1 = m)}{m}; \\ . & .. & . \\ . & .. & . \\ \overline{W}(a_{\max}, b_1) = \frac{\sum_{j=1}^{m} W_j(a_{\max}, b_1)}{m}; & .. & \overline{W}(a_{\max}, b = m) = \frac{\sum_{j=1}^{m} W_j(a_{\max}, b_1 = m)}{m}. \end{pmatrix}. \tag{1.3}$$

Examples of the application of this wavelet analysis to the maps of azimuth and ellipticity of polarization of microscopic images of polycrystalline films of blood plasma are given in [73, 74] (Figs. 1.2 and 1.3).

From the data given in [73, 74], it follows that by means of wavelet analysis of the distribution of azimuth and polarization ellipticity of microscopic images of polycrystalline films of human blood plasma, a relationship was established between changes

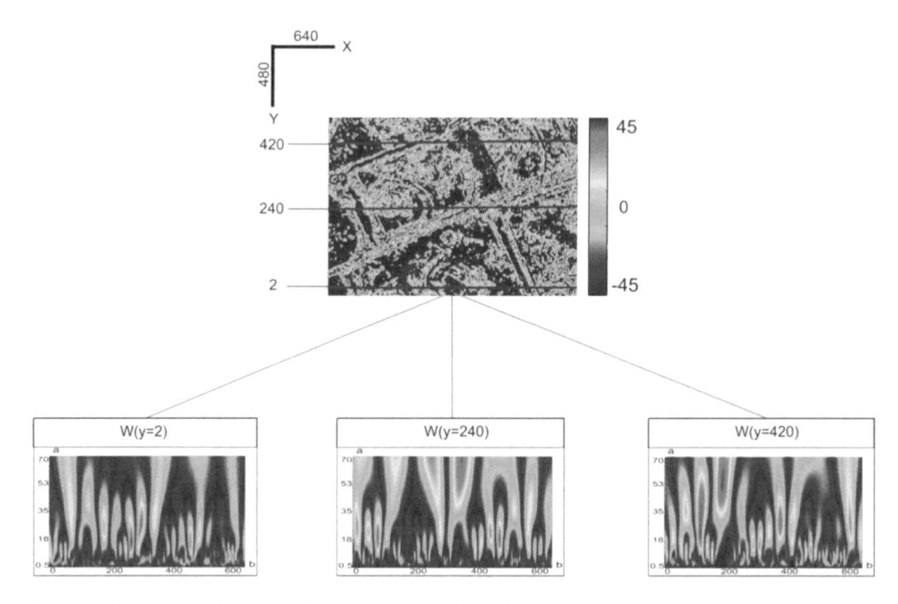


**Fig. 1.2** Wavelet-coefficients $W_{a,b}(m \times n)$ of the polarization map of the azimuths $\alpha(x, y)$ of human blood plasma with an acute inflammatory process for various lines $y = \left(k1; \ . \ km\right)$

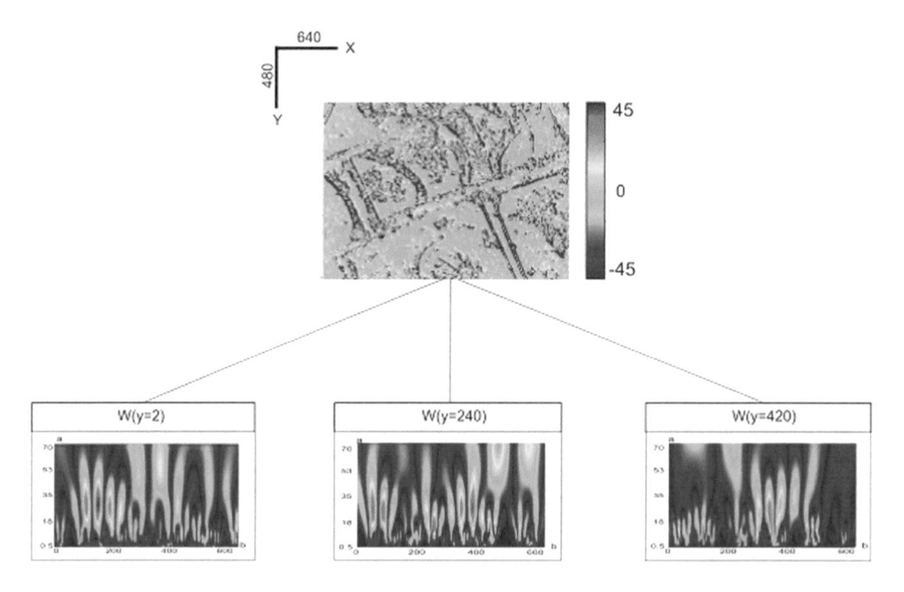


**Fig. 1.3** Wavelet-coefficients $W_{a,b}(m \times n)$ of the polarization map of the ellipticity $\beta(x, y)$ of human blood plasma with an acute inflammatory process for various lines $y = \left(k1; \ . \ km\right)$

in the statistical moments characterizing the set of wavelet coefficients and the distribution of directions of the optical axes and birefringence of albumin amino acid crystals and globulin of an optical anisotropic polycrystalline network at different scales of its geometric dimensions. As a result, statistical criteria for differentiating various scale levels of polycrystalline networks with distributions of wavelet coefficients of azimuth and ellipticity of polarization of microscopic images of blood plasma were established for the first time [73, 74]. On this basis, differentiation of the type of pathology of the human body—heart failure and acute inflammatory processes—has been established.

The wavelet analysis revealed sensitivity to changes in the optical anisotropy of polycrystalline networks of biological fluids and tissues at different scales of their structural organization. This fact is of fundamental importance in comparison with polarization mapping methods. The fact is that, from a medical point of view, various types of pathology appear at an early stage when looking at small-scale morphological structures. Therefore, the different-scale differentiation of the manifestations of optical anisotropy has a clearly defined applied aspect. On the other hand, the wavelet analysis method has a fundamental basis—it is a coordinate-localized correlation (convolution of the "window" of the wavelet function and coordinate signal features) analysis of the distribution of polarization states of microscopic images and elements of the Mueller matrix of biological layers. Based on this, we can discuss the correlation improvement of the polarization or Mueller-matrix mapping technique. However, such a correlation approach is purely analytical, carrying not only information about the optical anisotropy of multiscale structures. The main disadvantages of this method include the significant dependence of the results on the choice of the wavelet function. In this sense, another experimental method, which is based on consistent (correlation) spatial-frequency filtering of polarized-inhomogeneous fields of scattered radiation, is more attractive and reliable.

## 1.4 Fourier Analysis of Polarization-Inhomogeneous Fields

As we have already noted, for the development of laser-polarimetric diagnostic methods, the task of differentiating optical anisotropy at different scales of the geometric dimensions of biological objects is important. One of the possible directions was the technique of coordinated spatial-frequency filtering described in [12–14, 75]. Here, spatial-frequency filtering of polarization-inhomogeneous microscopic images of polycrystalline networks of biological layers is applied. This technique is based on three stages: "direct Fourier transform," "spatial frequency filtering," and "inverse Fourier transform" of microscopic images of the biological layer. This direction of laser polarimetry is called Fourier–Stokes polarimetry.

The data obtained in [75] revealed the possibility of differentiating the low- and high-frequency components of the distributions of polarization azimuths in the image of a grid of birefringent cylinders. It is established that the statistical moments of the third and fourth orders, characterizing the distribution of the large-scale component of the map of polarization azimuths, vary from 2 to 3.5 times. For statistical moments of the second to fourth orders characterizing the distribution of the polarization azimuth of the small-scale component of the polarization map, the differences between the values of such parameters are in the range from 1.5 to 4 times. The information obtained in this way became the basis for the experimental differentiation of polycrystalline films of blood plasma from a healthy person and a patient with prostate cancer. Diagnostically applicable parameters with sensitivity to the oncological state of the human prostate have been established. These are the statistical moments of the third and fourth orders characterizing the distribution of polarization azimuth values of the large-scale component of the polarization map. The differences between their values are $7(M_3) - 10(M_4)$ times. For the small-scale component of the azimuth map, differences between the values of statistical moments of the second- to fourth-order statistical moments characterizing the polarization distributions from 2 ($M_2$, $M_3$) to 10 ($M_4$) times are revealed.

In parallel with the analyzed directions of the correlation (wavelet and Fourier) analysis of polarization-inhomogeneous images, traditional correlation approaches, which formed the basis of polarization correlometry, became widespread.

## 1.5 Correlation Approaches to the Analysis of Polarization-Inhomogeneous Fields

Consider the theoretical foundations of polarization correlometry. The first attempt to analyze polarization-inhomogeneous fields was a generalization of the coherence matrix [1, 12] with a two-point polarization coherence matrix [22]. Such an approach was developed in [23] by generalizing the theory of coherence and polarization. In order to generalize the coherence matrix, a new parameter was proposed in [24]—the degree of coherence. It characterizes the correlation similarity of the electric field at two points. In parallel with this, the degree of polarization was generalized in [25], which made it possible to envisage spatial changes in the polarization state along the beam propagation direction. Finally, in [26] all the correlation and polarization approaches to the description of optical radiation fields were combined. A complex degree of mutual polarization (CDMP) is introduced here, which determines the correlation relationship between field points with different polarizations and intensities. Thus, it can be stated that the group of parameters [27–30] refers to the correlation basis for the description of scattered radiation fields by optically anisotropic layers. Based on the unified theory of coherence and polarization of random electromagnetic fields, it was possible to find the relationship between the parameters of partially polarized light and the correlation of unpolarized components [76, 77].

## 1.6 The Complex Degree of Mutual Polarization of Microscopic Images of Biological Tissues

In a series of works [44–48, 76–83], new approaches were developed for the correlation description of polarization-inhomogeneous laser object fields of biological tissues. Therein, methods are suggested for measuring the coordinate distributions of the modulus—the CDMP phase for diagnosing the structure of birefringent fibrillar networks. In [44–46, 77–83], two approaches to the analysis of the CDMP were proposed based on the determination of its modulus and phase. It is shown that the coordinate distribution of the CDMP modulus is determined by the distribution of directions of the optical axes of the ensemble of biological crystals. The CDMP phase is determined by phase shifts between the orthogonal components of the amplitude at different points in the object field.

The diagnostic capabilities of polarization correlometry of images of healthy and pathologically altered myometrium tissue are illustrated by the following series of results. It was found that the third-order statistical moment, which characterizes the distribution of the values of the CDMP $|V(x, y)|$ modulus of the microscopic image of a histological section of a benignly changed myometrial tissue, is three times less than the value of the third-order statistical moment calculated for the microscopic image of a histological section of a healthy myometrium tissue.

A larger range (up to ten times) of change is undergone by the kurtosis value characterizing the distribution of the values of the modulus and phase of CDMP microscopic images of histological sections of benignly altered tissue and healthy tissue.

The logical development of this technique was the development of a correlation approach to a direct description of the consistency of linear birefringence parameters of fibrillar networks based on the introduction of a parameter of the complex degree of mutual anisotropy (CDMA).

## 1.7 The Complex Degree of Mutual Anisotropy of Biological Tissues

In a series of works [76–83, 44–48], a new parameter, the complex degree of mutual anisotropy, was introduced to characterize polycrystalline networks:

$$W(r_{n+k}, r_n) = \frac{\left(\left(I^{(0)}(r_{n+k})I^{(0)}(r_n)\right)^{\frac{1}{2}} - \left(I^{(90)}(r_{n+k})I^{(90)}(r_n)\right)^{\frac{1}{2}}\right)^2}{I(r_{n+k})I(r_n)} + \frac{4\left(I^{(0)}(r_{n+k})I^{(90)}(r_{n+k})I^{(0)}(r_n)I^{(90)}(r_n)\right)^{\frac{1}{2}}\cos(\delta_{n+k}(r_{n+k}) - \delta_n(r_n))}{I(r_{n+k})I(r_n)} \tag{1.4}$$

Here, $I^{(0)}(r_i)$, $I^{(90)}(r_i)$ are the intensity values measured for orientation of the analyzer's transmission axis at angles $\Theta = 0°$ and $\Theta = 90°$ respectively.

This method was applied to identify scenarios of the formation of correlation contours (half-width line of the two-dimensional autocorrelation distribution function of the CDMA modulus values) of polycrystalline networks with different distributions of optical axis directions and phase modulation laws. The developed cross-correlation approach is illustrated by the data from computer modeling and experimental studies shown in Fig. 1.4.

It has been established that for a histological section of a biopsy of a benign tumor of the uterine wall, the correlation CDMA circuit is asymmetric due to the presence of birefringent fibril growth directions. For a histological section of a biopsy of a malignant tumor, the skewness of the correlation CDMA contour is halved due to the destruction of birefringent networks.

It should be noted that in order to obtain diagnostically suitable information, a prerequisite is the reproducibility of the experimental data of laser polarimetry and polarization correlometry. The analysis revealed that almost all studies in this direction are azimuthally dependent on the rotation of the sample plane relative to the direction of irradiation. Based on this, it is important not only to improve the methods of laser polarimetry by correlation approaches, but also to understand the conditions for azimuthal invariance of experimental data.

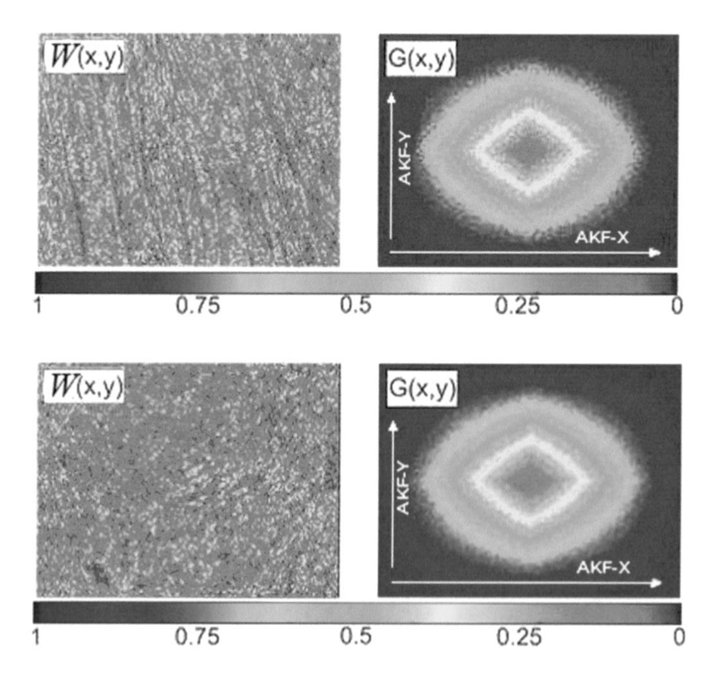


**Fig. 1.4** Coordinate CDMA distributions and their two-dimensional autocorrelation functions of optically thin histological sections of operationally removed benign (left column) and malignant (right column) uterine wall tumors

## 1.8 Azimuthal Polarization Invariants

The informational completeness and accessibility of the experimental measurement are determined by the fact that a number of practical Mueller-matrix techniques are used in biological and medical research [18, 19, 49–51, 31–35, 52]. Methods for two-dimensional Mueller-matrix mapping of biological tissues have been developed [37–39, 50, 56, 60–69]. However, not all matrix elements are convenient for specimen characterization. The reason for this is the dependence of the elements of the Muller matrix on the choice of the coordinate system. It is known that 12 out of the 16 matrix elements change when the sample rotates around the axis of irradiation. In addition, there is ambiguity in the relationship between the optical properties of the object of study and the set of elements of the Mueller matrices. Diagnostic information about the structure of a biological object is "encrypted" in the Mueller matrix. To overcome this ambiguity, the method of polar decomposition of Mueller matrices was developed [31, 41, 51, 55, 57, 59, 61, 71]. It is shown therein that the following elements and combinations of elements of the Mueller matrix remain constant

$$m_{11}, m_{14}, m_{41}, m_{44}, m_{22} + m_{33}, m_{23} - m_{32}, \tag{1.5}$$

and also the length of the following vectors is conserved

$$\mathbf{a}_{\mathrm{H}} = \begin{pmatrix} m_{12} \\ m_{13} \end{pmatrix}, \mathbf{a}_{\mathrm{V}} = \begin{pmatrix} m_{21} \\ m_{31} \end{pmatrix}, \mathbf{b}_{\mathrm{H}} = \begin{pmatrix} m_{42} \\ m_{43} \end{pmatrix}, \mathbf{b}_{\mathrm{V}} = \begin{pmatrix} m_{24} \\ m_{34} \end{pmatrix}, \mathbf{g} = \begin{pmatrix} m_{22} - m_{33} \\ m_{23} + m_{32} \end{pmatrix}, \tag{1.6}$$

Therefore, in the future, when developing and systematizing polarization correlometry methods, we will use the set of indicated Mueller-matrix invariants (MMI).

## References

1. V. Tuchin, L. Wang, D. Zimnjakov, *Optical Polarization in Biomedical Applications* (Springer, New York, USA, 2006).
2. R. Chipman, *Polarimetry.* in ed. by M. Bass. Handbook of Optics: Vol I—Geometrical and Physical Optics, Polarized Light, Components and Instruments (McGraw-Hill Professional, New York, 2010) , pp. 22.1–22.37
3. N. Ghosh, M. Wood, A. Vitkin, in *Polarized Light Assessment of Complex Turbid Media such as Biological Tissues Via Mueller Matrix Decomposition*, ed. by V. Tuchin. Handbook of Photonics for Biomedical Science (CRC Press, Taylor & Francis Group, London, 2010), pp. 253–282
4. S. Jacques, Polarized Light Imaging of Biological Tissues, in *Handbook of Biomedical Optics.* ed. by D. Boas, C. Pitris, N. Ramanujam (CRC Press, Boca Raton, London, New York, 2011), pp. 649–669
5. N. Ghosh, Tissue polarimetry: concepts, challenges, applications, and outlook. J. Biomed. Opt. **16**(11), 110801 (2011)

6. M. Swami, H. Patel, P. Gupta, Conversion of 3×3 Mueller matrix to 4×4 Mueller matrix for non-depolarizing samples. Opt. Commun. **286**, 18–22 (2013)
7. D. Layden, N. Ghosh, A. Vitkin, in *Quantitative Polarimetry for Tissue Characterization and Diagnosis*, ed. by R. Wang, V. Tuchin. Advanced Biophotonics: Tissue Optical Sectioning (CRC Press, Taylor & Francis Group, Boca Raton, London, New York, 2013) , pp. 73–108
8. T. Vo-Dinh, *Biomedical Photonics Handbook*, 3 vol. set, 2nd edn. (CRC Press, Boca Raton, 2014)
9. A. Vitkin, N. Ghosh, A. Martino, Tissue Polarimetry, in *Photonics: Scientific Foundations, Technology and Applications*, 4th edn., ed. by D. Andrews (Wiley, Hoboken, New Jersey, 2015), pp. 239–321
10. V. Tuchin, *Tissue Optics: Light Scattering Methods and Instruments for Medical Diagnosis*, 2nd edn. (SPIE Press, Bellingham, Washington, USA, 2007).
11. W. Bickel, W. Bailey, Stokes vectors, Mueller matrices, and polarized scattered light. Am. J. Phys. **53**(5), 468–478 (1985)
12. A. Doronin, C. Macdonald, I. Meglinski, Propagation of coherent polarized light in turbid highly scattering medium. J. Biomed. Opt. **19**(2), 025005 (2014)
13. A. Doronin, A. Radosevich, V. Backman, I. Meglinski, Two electric field Monte Carlo models of coherent backscattering of polarized light. J. Opt. Soc. America A **31**(11), 2394 (2014)
14. A. Ushenko, V. Pishak, in *Laser Polarimetry of Biological Tissue: Principles and Applications*, ed. by V. Tuchin. Handbook of Coherent-Domain Optical Methods: Biomedical Diagnostics (Environmental and Material Science, 2004) , pp. 93–138
15. P. Li, H.R. Lee, S. Chandel, C. Lotz, F. Kai Groeber-Becker, S. Dembski, R. Ossikovski, H. Ma, T. Novikova, Analysis of tissue microstructure with Mueller microscopy: logarithmic decomposition and Monte Carlo modeling. J. Biomed. Opt. **25**(1), (2020)
16. O. Angelsky, A. Ushenko, Y. Ushenko, V. Pishak, A. Peresunko, in *Statistical, Correlation and Topological Approaches in Diagnostics of the Structure and Physiological State of Birefringent Biological Tissues*. Handbook of Photonics for Biomedical Science (2010), pp. 283–322
17. Y. Ushenko, T. Boychuk, V. Bachynsky, O. Mincer, in *Diagnostics of Structure and Physiological State of Birefringent Biological Tissues: Statistical, Correlation and Topological Approaches*, ed. by V. Tuchin. Handbook of Coherent-Domain Optical Methods (Springer Science+Business Media, 2013)
18. O. Angelsky, A. Ushenko, Y. Ushenko, Investigation of the correlation structure of biological tissue polarization images during the diagnostics of their oncological changes. Phys. Med. Biol. **50**(20), 4811–4822 (2005)
19. V. Ushenko, O. Dubolazov, A. Karachevtsev, Two wavelength Mueller matrix reconstruction of blood plasma films polycrystalline structure in diagnostics of breast cancer. Appl. Opt. **53**(10), B128 (2016)
20. A. Ushenko, A. Sdobnov, A. Dubolazov, M. Grytsiuk, Y. Ushenko, A. Bykov, I. Meglinski, Stokes-correlometry analysis of biological tissues with polycrystalline structure. IEEE J. Sel. Top. Quantum Electron. **25**(1), 8438957 (2019)
21. O. Vanchulyak, O. Ushenko, V. Zhytaryuk, V. Dvorjak, O. Pavlyukovich, O. Dubolazov, N. Pavlyukovich, N.P. Penteleichuk, Stokes-correlometry of polycrystalline films of biological fluids in the early diagnostics of system pathologies. Proc. SPIE—Int. Soc. Opt. Eng. **11105**, 1110519 (2019)
22. E. Wolf, Unified theory of coherence and polarization of random electromagnetic beams. Phys. Lett. A **312**, 263–267 (2003)
23. J. Tervo, T. Setala, A. Friberg, Degree of coherence for electromagnetic. Opt. Express **11**, 1137–1143 (2003)
24. J.M. Movilla, G. Piquero, R. Martínez-Herrero, P.M. Mejías, Parametric characterization of non-uniformly polarized. Opt. Commun. **149**, 230–234 (1998)
25. J. Ellis, A. Dogariu, Complex degree of mutual polarization. Opt. Lett. **29**, 536–538 (2004)
26. C. Mujat, A. Dogariu, Statistics of partially coherent beams: a numerical analysis. J. Opt. Soc. Am. A **21**(6), 1000–1003 (2004)

27. F. Gori, Matrix treatment for partially polarized, partially coherent beams. Opt. Lett. **23**, 241–243 (1998)
28. E. Wolf, Significance and measurability of the phase of a spatially coherent optical field. Opt. Lett. **28**, 5–6 (2003)
29. M. Mujat, A. Dogariu, Polarimetric and spectral changes in random electromagnetic fields. Opt. Lett. **28**, 2153–2155 (2003)
30. J. Ellis, A. Dogariu, S. Ponomarenko, E. Wolf, Interferometric measurement of the degree of polarization and control of the contrast of intensity fluctuations. Opt. Lett. **29**, 1536–1538 (2004)
31. S. Manhas, M.K. Swami, P. Buddhiwant, N. Ghosh, P.K. Gupta, K. Singh, Mueller matrix approach for determination of optical rotation in chiral turbid media in backscattering geometry. Opt. Exp. **14**, 190–202 (2006)
32. Y. Deng, S. Zeng, Q. Lu, Q. Luo, Characterization of backscattering Mueller matrix patterns of highly scattering media with triple scattering assumption. Opt. Exp. **15**, 9672–9680 (2007)
33. S.Y. Lu, R.A. Chipman, Interpretation of Mueller matrices based on polar decomposition. J. Opt. Soc. Am. A **13**, 1106–1113 (1996)
34. Y. Guo, N. Zeng, H. He, T. Yun, E. Du, R. Liao, H. Ma, A study on forward scattering Mueller matrix decomposition in anisotropic medium. Opt. Exp. **21**, 18361–18370 (2013)
35. A. Pierangelo, S. Manhas, A. Benali, C. Fallet, J.L. Totobenazara, M.R. Antonelli, P. Validire, Multispectral Mueller polarimetric imaging detecting residual cancer and cancer regression after neoadjuvant treatment for colorectal carcinomas. J. Biomed. Opt. **18**, 046014 (2013)
36. V. Devlaminck, Physical model of differential Mueller matrix for depolarizing uniform media. J. Opt. Soc. America A **30**(11), 2196 (2013)
37. Y.A. Ushenko, G.D. Koval, A.G. Ushenko, O.V. Dubolazov, V.A. Ushenko, O.Y. Novakovskaia, Mueller-matrix of laser-induced autofluorescence of polycrystalline films of dried peritoneal fluid in diagnostics of endometriosis. J. Biomed. Opt. **21** (7), 071116 (2016)
38. M. Borovkova, M. Peyvasteh, O. Dubolazov, Y. Ushenko, V. Ushenko, A. Bykov, S. Deby, J. Rehbinder, T. Novikova, I. Meglinski, Complementary analysis of Mueller-matrix images of optically anisotropic highly scattering biological tissues. J. Eur. Opt. Soc. **14**(1), 20 (2018)
39. V. Ushenko, A. Sdobnov, A. Syvokorovskaya, A. Dubolazov, O. Vanchulyak, A. Ushenko, Y. Ushenko, M. Gorsky, M. Sidor, A. Bykov, I. Meglinski, 3D Mueller-matrix diffusive tomography of polycrystalline blood films for cancer diagnosis. Photonics **5**(4), 54 (2018)
40. L. Trifonyuk, W. Baranowski, V. Ushenko, O. Olar, A. Dubolazov, Y. Ushenko, B. Bodnar, O. Vanchulyak, L. Kushnerik, M. Sakhnovskiy, 2D-Mueller-matrix tomography of optically anisotropic polycrystalline networks of biological tissues histological sections. Opto-Electron. Rev. **26**(3), 252–259 (2018)
41. V.A. Ushenko, A.V. Dubolazov, L.Y. Pidkamin, M.Y. Sakchnovsky, A.B. Bodnar, Y.A. Ushenko, A.G. Ushenko, A. Bykov, I. Meglinski, Mapping of polycrystalline films of biological fluids utilizing the Jones-matrix formalism. Laser Phys. **28**(2), 025602 (2018)
42. V.A. Ushenko, A.Y. Sdobnov, W.D. Mishalov, A.V. Dubolazov, O.V. Olar, V.T. Bachinskyi, A.G. Ushenko, Y.A. Ushenko, O.Y. Wanchuliak, I. Meglinski, Biomedical applications of Jones-matrix tomography to polycrystalline films of biological fluids. J. Innovative Opt. Health Sci. **12**(6), 1950017 (2019)
43. M. Borovkova, L. Trifonyuk, V. Ushenko, O. Dubolazov, O. Vanchulyak, G. Bodnar, Y. Ushenko, O. Olar, O. Ushenko, M. Sakhnovskiy, A. Bykov, I. Meglinski, Mueller-matrix-based polarization imaging and quantitative assessment of optically anisotropic polycrystalline networks. PLoS ONE **14**(5), e0214494 (2019)
44. O.Y. Novakovska, Polarization correlometry of characteristic states of Muller-matrix images of phase-inhomogeneous biological layers. Semicond. Phys. Quantum Electron. Optoelectron. **15**(3), 230–237 (2012)
45. A.G. Ushenko, Y.A. Ushenko, Y.Y. Tomka, O.V. Dubolazov, O.Y. Telenga, V.I. Istratiy, A.O. Karachevtsev, The interconnection between the coordinate distribution of Muller-matrixes images characteristics values of biological liquid crystals net and the pathological changes of human tissues : 12–16 July 2010, 9th Euro-American Workshop on Information Optics. Helsinki, Finland (2010)

46. Y.A. Ushenko, O.V. Dubolazov, O.Y. Telenga, A.P. Angelsky, A.O. Karachevtcev, V. Balanetska, Complex degree of mutual anisotropy of biological liquid crystals net. Proc. SPIE **8087**, 80872Q (2011)
47. O.V. Angelsky, Y.A. Ushenko, V.O. Balanetska, The degree of mutual anisotropy of biological liquids polycrystalline nets as a parameter in diagnostics and differentiations of hominal inflammatory processes. Proc. SPIE **8338**, 83380S (2011)
48. Y.A. Ushenko, A.V. Dubolazov, A.O. Karachevtcev, N.I. Zabolotna, Complex degree of mutual anisotropy in diagnostics of biological tissues physiological changes. Proc. SPIE **8134**, 81340O (2011)
49. Y. Ushenko, G. Koval, A. Ushenko, O. Dubolazov, V. Ushenko, O. Novakovskaia, Mueller-matrix of laser-induced autofluorescence of polycrystalline films of dried peritoneal fluid in diagnostics of endometriosis. J. Biomed. Opt. **21**(7), 071116 (2016)
50. A. Ushenko, A. Dubolazov, V. Ushenko, O. Novakovskaya, Statistical analysis of polarization-inhomogeneous Fourier spectra of laser radiation scattered by human skin in the tasks of differentiation of benign and malignant formations. J. Biomed. Opt. **21**(7), 071110 (2016)
51. V. Ushenko, N. Pavlyukovich, L. Trifonyuk, Spatial-frequency azimuthally stable cartography of biological polycrystalline networks. Int. J. Opt. **2013**, 1–7 (2013)
52. V.P. Ungurian, O.I. Ivashchuk, V.O. Ushenko, Statistical analysis of polarizing maps of blood plasma laser images for the diagnostics of malignant formations. Proc. SPIE **8338**, 83381L (2011)
53. V.A. Ushenko, O.V. Dubolazov, A.O. Karachevtsev, Two wavelength Mueller matrix reconstruction of blood plasma films polycrystalline structure in diagnostics of breast cancer. Appl. Opt. **53**, B128–B139 (2014)
54. V.P. Prysyazhnyuk, Y.A. Ushenko, A.V. Dubolazov, A.G. Ushenko, V.A. Ushenko, Polarization-dependent laser autofluorescence of the polycrystalline networks of blood plasma films in the task of liver pathology differentiation. Appl. Opt. **55**, B126–B132 (2016)
55. V.A. Ushenko, M.S. Gavrylyak, Azimuthally invariant Mueller-matrix mapping of biological tissue in differential diagnosis of mechanisms protein molecules networks anisotropy. Proc. SPIE **8812**, 88120Y (2013)
56. V.A. Ushenko, M.P. Gorsky, Complex degree of mutual anisotropy of linear birefringence and optical activity of biological tissues in diagnostics of prostate cancer. Opt. Spectrosc. **115**, 290–297 (2013)
57. V.A. Ushenko, A.V. Dubolazov, Correlation and self similarity structure of polycrystalline network biological layers Mueller matrices images. Proc. SPIE **8856**, 88562D (2013)
58. V.O. Ushenko, Spatial-frequency polarization phasometry of biological polycrystalline networks. Opt. Mem. Neur. Netw. **22**, 56–64 (2013)
59. V.A. Ushenko, N.D. Pavlyukovich, L. Trifonyuk, Spatial-frequency azimuthally stable cartography of biological polycrystalline networks. Int. J. Opt. **683174**, 2013 (2013)
60. Y.A. Ushenko, Spatial-frequency Fourier polarimetry of the complex degree of mutual anisotropy of linear and circular birefringence in the diagnostics of oncological changes in morphological structure of biological tissues. Quantum. Electron. **42**, 727–732 (2012)
61. V.A. Ushenko, Complex degree of mutual coherence of biological liquids. Proc. SPIE **8882**, 88820V (2013)
62. Yu.A. Ushenko, Jones-matrix mapping of complex degree of mutual anisotropy of birefringent protein networks during the differentiation of myocardium necrotic changes. Appl. Opt. **55**, B113–B119 (2016)
63. L. Cassidy, Basic concepts of statistical analysis for surgical research. J. Surg. Res. **128**(2), 199–206 (2005)
64. C.S. Davis, *Statistical Methods of the Analysis of Repeated Measurements* (Springer, New York, 2002).
65. A. Petrie, C. Sabin, *Medical Statistics at a Glance* (Wiley-Blackwell, Chichester, UK, 2009).
66. A.G. Ushenko, O.V. Dubolazov, V.A. Ushenko, O.Y. Novakovskaya, O.V. Olar, Fourier polarimetry of human skin in the tasks of differentiation of benign and malignant formations. Appl. Opt. **55**(12), B56–B60 (2016)

67. Y.A. Ushenko, V.T. Bachynsky, O.Y. Vanchulyak, A.V. Dubolazov, M.S. Garazdyuk, V.A. Ushenko, Jones-matrix mapping of complex degree of mutual anisotropy of birefringent protein networks during the differentiation of myocardium necrotic changes. Appl. Opt. **55**(12), B113-B119 (2016)
68. A.V. Dubolazov, N.V. Pashkovskaya, Y.A. Ushenko, Y.F. Marchuk, V.A. Ushenko, O.Y. Novakovskaya, Birefringence images of polycrystalline films of human urine in early diagnostics of kidney pathology. Appl. Opt. **55**(12), B85–B90 (2016)
69. M.S. Garazdyuk, V.T. Bachinskyi, O.Y. Vanchulyak, A.G. Ushenko, O.V. Dubolazov, M.P. Gorsky, Polarization-phase images of liquor polycrystalline films in determining time of death. Appl. Opt. **55**(12), B67-B71 (2016)
70. A.V. Dubolazov, O.V. Olar, L.Y. Pidkamin, A.D. Arkhelyuk, A.V. Motrich, V.T. Bachinskiy, O.V. Pavliukovich, N. Pavliukovich, Differential components of Muller matrix partially depolarizing biological tissues in the diagnosis of pathological and necrotic changes. Proc. SPIE **11087**, 1108713 (2019)
71. O. Ushenko, V. Zhytaryuk, V. Dvorjak, I.V. Martsenyak, O. Dubolazov, B.G. Bodnar, O.Y. Vanchulyak, S. Foglinskiy, Multifunctional polarization mapping system of networks of biological crystals in the diagnostics of pathological and necrotic changes of human organs. Proc. SPIE **11087**, 110870S (2019)
72. O. Pavlyukovich, N. Pavlyukovich, Y. Ushenko, O. Galochkin, M. Sakhnovskiy, M. Kovalchuk, A. Dovgun, S. Golub, O. Dubolazov, Fractal analysis of patterns for birefringence biological tissues in the diagnostics of pathological and necrotic states. Proc. SPIE **11105**, 1110518 (2019)
73. Y.Y. Tomka, Wavelet analysis of biological tissue's Mueller-matrix images. Proc. SPIE **7008**, 700823 (2008)
74. O.V. Dubolazov, Y.O. Ushenko, Y.Y. Tomka, O.G. Pridiy, A.V. Motrich, I.Z. Misevitch, V.V. Istratiy, Wavelet analysis for Mueller matrix images of biological crystal networks. Semicond. Phys. Quantum Electron. Optoelectron. **12**(4), 391–398 (2009)
75. A.O. Karachevtsev, Fourier Stokes-polarimetry of biological layers polycrystalline networks. Semicond. Phys. Quantum Electron. Optoelectron. **15**(3), 252–268 (2012)
76. O. Angelsky, A. Ushenko, Y. Ushenko, Complex degree of mutual polarization of biological tissue coherent images for the diagnostics of their physiological state. J. Biomed. Opt. **10**(6), 060502 (2005)
77. Y. Ushenko, Complex degree of mutual polarization of Biotissue's Speckle-images. Ukr. J. Phys Opt. **6**(3), 104–113 (2005)
78. S.B. Yermolenko, C.Y. Zenkova, A.-P. Angelskiy, Polarization manifestations of correlation (intrinsic coherence) of optical fields. Appl. Opt. **47**(32) (2008)
79. O.V. Angelsky, A.G. Ushenko, A.O. Angelskaya, Y.A. Ushenko, Correlation- and singular-optical approaches in diagnostics of polarization inhomogeneity of coherent opical fields from biological tissues. Ukr. J. Phys. Opt. **8**(2), 106–123 (2007)
80. O.V. Angelsky, A.G. Ushenko, A.O. Angelskaya, Y.A. Ushenko, Polarization correlometry of polarization singularities of biological tissues object fields. Proc. SPIE **6616**, 1–9 (2007)
81. Y.O. Ushenko, Y.Y. Tomka, O.I. Telenga, I.Z. Misevitch, V.V. Istratiy, Complex degree of mutual anisotropy of biological liquid crystals nets. Opt. Eng. **50**, 039001 (2011)
82. Y.A. Ushenko, O.I. Telenga, A.P. Peresunko, O.K. Numan, New parameter for describing and analyzing the optical-anisotropic properties of biological tissues. J. Innov. Opt. Health Sci. **4**(4), 463–475 (2011)
83. A.V. Dubolazov, O.Y. Telenha, V.A. Ushenko, M.I. Sydor, Characteristic values of Mueller-matrixes images of biological liquid crystals net for diagnostics of human tissues anisotropy. Proc. SPIE **8338**, 83380Z (2011)

# Chapter 2
# Materials and Methods

## 2.1 Azimuthally Invariant Polarization Mapping System

This section discusses the optical scheme and architecture of an azimuthally invariant polarization mapping system of the distributions of azimuth and polarization ellipticity microscopic images of histological sections of biological tissues.

Figure 2.1 presents an optical scheme of a classical Stokes polarimeter, which is modified for azimuthally invariant polarization mapping of microscopic images of biological layers.

Samples of native histological sections of biological tissues are irradiated with a parallel ($\varnothing = 2 \times 10^3$ μm) low-intensity ($W = 5.0$ mW) He–Ne laser beam ($\lambda = 0.6328$). The polarization state of the probe beam was formed using a quarter-wave plate 4, (Achromatic True Zero-Order Waveplate) and polarizer 3 (B + W Kaesemann XS-Pro Polarizer MRC Nano). In order to obtain the conditions of azimuthal invariance of polarization mapping, the fast axis of the quarter-wave plate 4 was oriented at an angle relative to the transmission axis of the polarizer 3. As a result, a right circularly polarized irradiating beam was formed (shown schematically in fragment 11). Under such conditions, the rotations of the plane of the sample of the histological Sect. 2.6 relative to the direction of irradiation are azimuthally invariant.

Analytically, this fact illustrates the solution of the following vector-parametric equation

$$S^* = \{F\}S^{\otimes} = \begin{Vmatrix} 1 & 0 & 0 & 0 \\ 0 & f_{22} & f_{23} & f_{24} \\ 0 & f_{32} & f_{33} & f_{34} \\ 0 & f_{42} & f_{43} & f_{44} \end{Vmatrix} \times \begin{pmatrix} 1 \\ 0 \\ 0 \\ 1 \end{pmatrix} = \begin{pmatrix} 1 \\ f_{24} \\ f_{34} \\ f_{44} \end{pmatrix}. \tag{2.1}$$



I. Meglinski et al., *Shedding the Polarized Light on Biological Tissues*,
SpringerBriefs in Applied Sciences and Technology,
https://doi.org/10.1007/978-981-10-4047-4_2

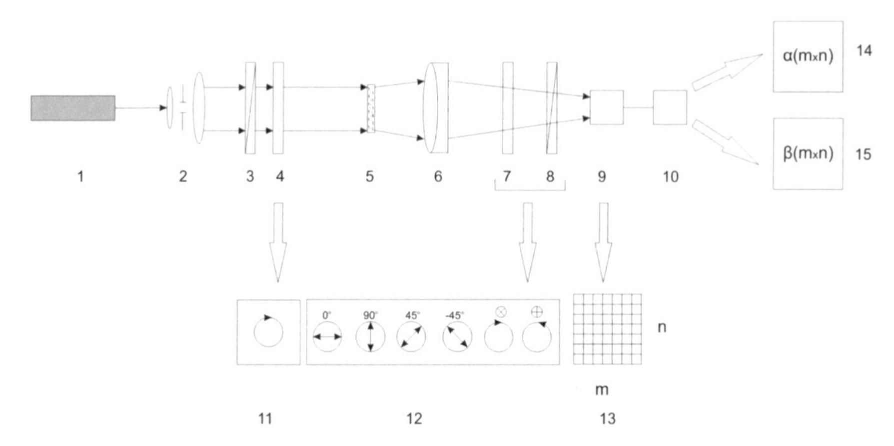


**Fig. 2.1** Optical scheme for azimuthally invariant polarization mapping of microscopic images of biological layers. Explanation in the text.

Here, $S^{\otimes}$ is the Stokes vector of the circularly polarized irradiating beam; $S^{*}$ is the Stokes vector of the object beam; $\{F\}$ is the Mueller matrix of the biological layer; $f_{ik}$ are the elements of the Mueller matrix.

It can be shown that for a birefringent crystal (Chap. 1, relation (1.1)), the azimuth $\alpha^{*}$ and ellipticity $\beta^{*}$ of the polarization of the object beam are determined by the relations

$$\alpha^{*}(\delta) = 0,5\mathrm{arctg}\left(\frac{f_{34}}{f_{24}}\right) = 0,5\mathrm{arctg}(ctg\delta); \tag{2.2}$$

$$\beta^{*}(\delta) = 0,\arcsin(f_{44}) = 0,5\arcsin(\cos\delta). \tag{2.3}$$

It is easy to see that the polarization parameters $\alpha^{*}$, $\beta^{*}$ exclusively depend on the magnitude of the azimuthally invariant phase shift $\delta$ between the orthogonal components of the amplitude of the laser wave.

Polarization images of a histological section of biological tissue 6 using a polarizing microlens 7 (Nikon CFI Achromat P, focal length 30 mm, numerical aperture 0.1, magnification 4×) were projected onto the plane of the photosensitive area ($m \times n = 1280 \times 960$ pixels) of the CCD camera 10 (The Imaging Source DMK 41AU02.AS, monochrome 1/2″ CCD, Sony ICX205AL (progressive scan); resolution—1280 × 960; photosensitive area—7600 × 6200 km; sensitivity—0.05 lx; dynamic range— bit, SNR—9 bit)—schematically indicated by 13.

Image analysis of histological sections of biological tissues 5 was carried out using a polarizer 8 and a quarter-wave plate 7. A schematic functional sequence of polarizing filtering of the object laser radiation is presented in fragment 12.

The use of statistical analysis of discrete two-dimensional azimuthally invariant azimuth and ellipticity arrays at points of microscopic images of biological layers (Fig. 2.2) allows the expansion of the functionality of registration systems of classical

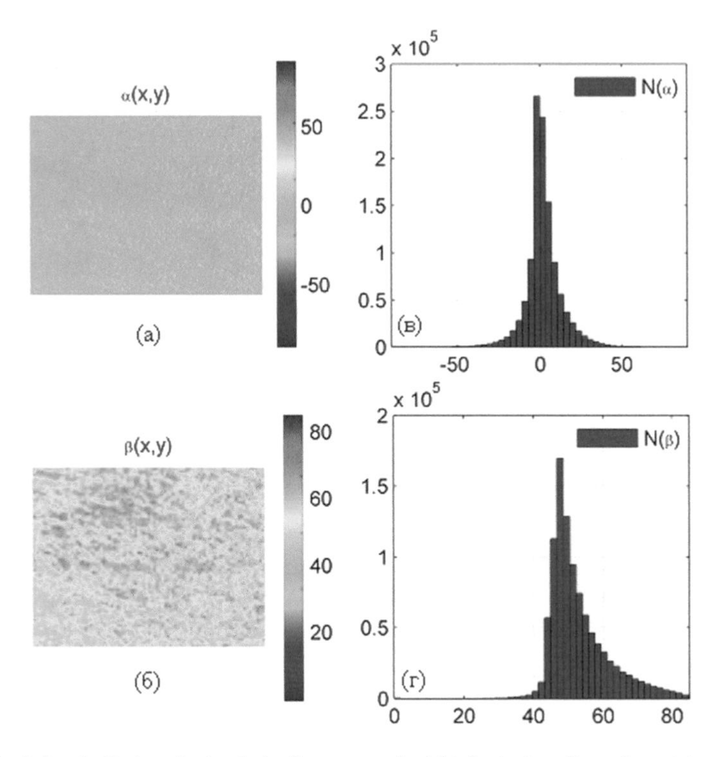


**Fig. 2.2** Azimuthally invariant polarization maps of a histological section of prostate adenoma tissue and histograms of the distribution of the azimuth and polarization elliptic values

microscopic images of biological layers [1–11] and obtaining new data on the fourth-order statistical moments characterizing the distributions $\alpha(m \times n)$ and $\beta(m \times n)$.

## 2.2 The System of Azimuthally Invariant Mueller-Matrix Mapping of Biological Layers

Figure 2.3 presents an optical scheme of a classical Stokes polarimeter, which is modified for azimuthally invariant Mueller-matrix mapping of microscopic images of histological sections of biological tissues.

The biological layer 6 was sequentially probed with a laser beam 1 with the following states of polarization: linear with azimuths of $0°, 90°, 45°$ and right-handed circular ($\otimes$), formed by a polarizing irradiator consisting of quarter-wave plates 3, 5 and polarizer 4, shown in 12.

Two groups of elements were chosen as suitable Mueller-matrix invariants, the physical properties of which were studied in detail in [12–14]:

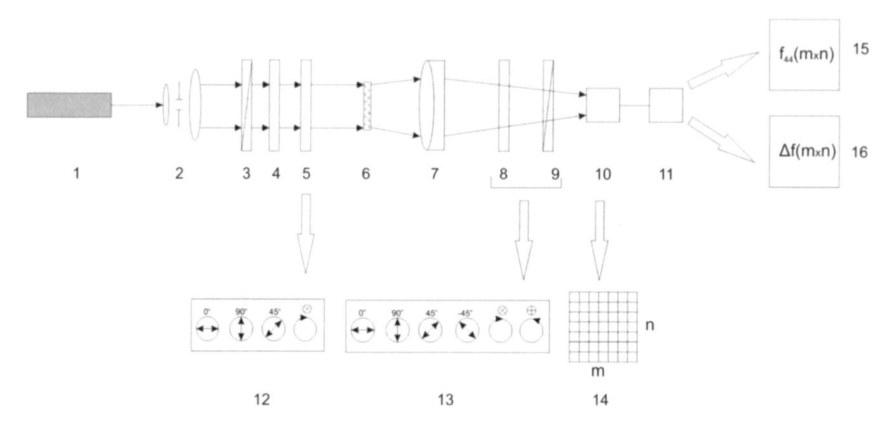


**Fig. 2.3** Optical design of an azimuthally invariant Mueller-matrix mapping of biological layers. Explanation in the text

- $f_{44}$—characterizes the "macroscopic" phase anisotropy of fibrillar networks (linear birefringence)—fragment 15;
- $\Delta f = \frac{f_{23} - f_{32}}{f_{22} + f_{33}}$—characterizes the "microscopic" phase anisotropy (circular birefringence) of the polypeptide chains that form the fibrils—fragment 16.

The calculation of these Mueller-matrix elements was carried out using a PC 11 according to the following algorithm [12]

$$\begin{aligned} &f_{22} = 0.5(S_2^0 - S_2^{90}); && f_{32} = 0.5(S_3^0 - S_3^{90}); && f_{41} = 0.5(S_4^0 + S_4^{90}); \\ &f_{23} = S_2^{45} - f_{21}; && f_{33} = S_3^{45} - f_{31}; && f_{44} = S_4^{\otimes} - f_{41}. \end{aligned} \tag{2.4}$$

Here, $S_{i=2;3;4}^{0;45;90;\otimes}$ are the Stokes vector parameters at the points of the digital image of histological Sect. 2.6, measured by a digital camera 10 with a two-dimensional set $(m \times n)$ of pixels 14 for a series of linearly (0°; 45°; 90°) and right-circular (⊗) polarized probe laser beams

$$\begin{aligned} S_{i=1}^{0;45;90;\otimes} &= I_0^{0;45;90;\otimes} + I_{90}^{0;45;90;\otimes}; \\ S_{i=2}^{0;45;90;\otimes} &= I_0^{0;45;90;\otimes} - I_{90}^{0;45;90;\otimes}; \\ S_{i=3}^{0;45;90;\otimes} &= I_{45}^{0;45;90;\otimes} - I_{135}^{0;45;90;\otimes}; \\ S_{i=4}^{0;45;90;\otimes} &= I_{\otimes}^{0;45;90;\otimes} + I_{\oplus}^{0;45;90;\otimes}. \end{aligned} \tag{2.5}$$

Here, $I_{0;45;90;135;\otimes;\oplus}$ are the intensities of the radiation transmitted by the linear polarizer 9 with the azimuth of rotation 0°; 45°; 90°; 135°, and the right- (⊗) and left-handed (⊕) circularly polarized filter 8 and 9 (see Fig. 2.4). The functional sequence of the polarization filtering algorithm is shown in fragment 13.

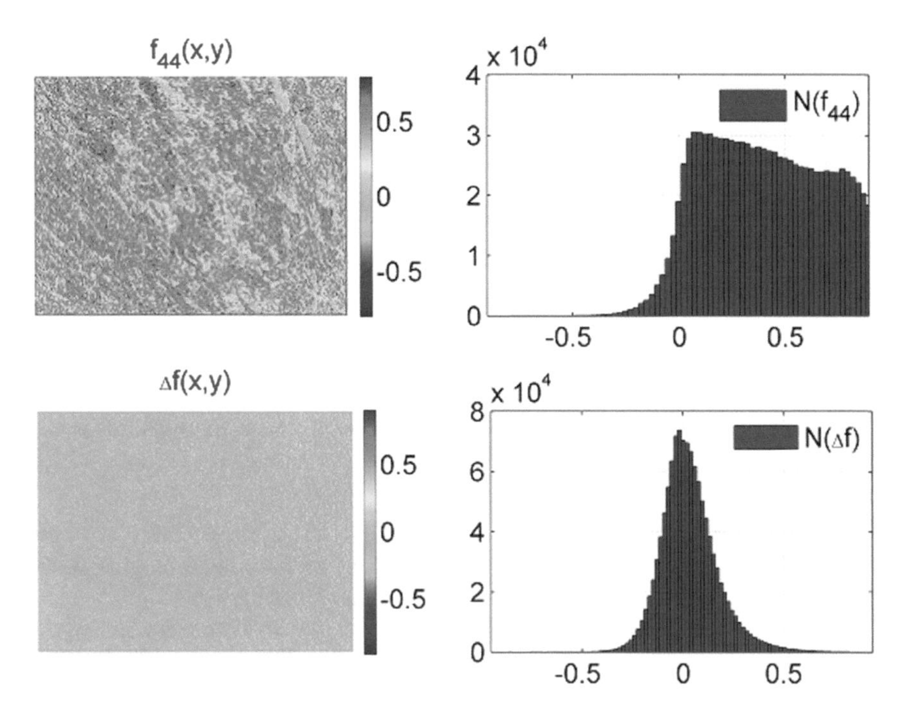


**Fig. 2.4** MMI histological section of prostate adenoma and histograms of the distribution of their random values

Figure 2.4 presents Mueller-matrix images of azimuthal invariants of the histological section of the layer of prostate adenoma and histograms of the distribution of random values of MMI.

The main result of the system of Mueller-matrix mapping is that it provides a new set of actions, which leads to the expansion of the functionality of the diagnostics of anisotropy of phase-inhomogeneous layers, and improves information content by implementing an azimuthally invariant process of measuring the coordinate distributions of the elements of the Mueller matrix.

## 2.3 CDMP Mapping System

Figure 2.5 presents the optical scheme of the CDMP mapping of microscopic images of biological layers.

The procedure for experimental measurements of the distribution of the values of the CDMP modulus from microscopic images of the biological layer is as follows:

1. Using a linear polarizer 3 and a quarter-wave plate 4, a circularly polarized irradiating beam is formed (see this chapter, paragraph 2.1)—fragment 11.

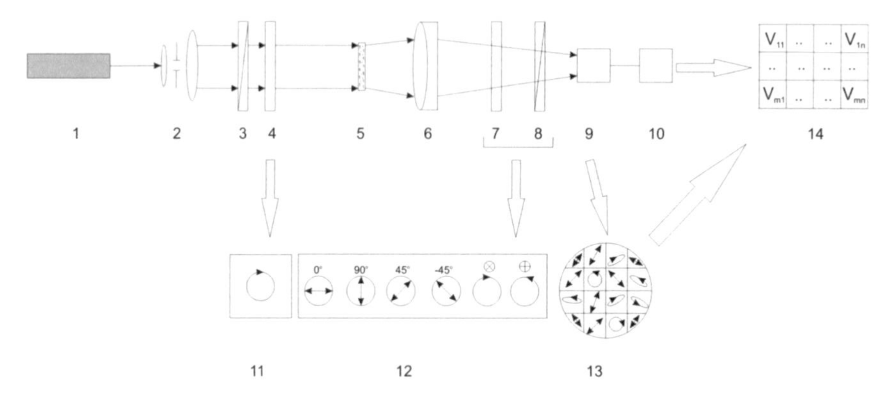


**Fig. 2.5** Optical design of CDMP mapping of microscopic images of biological layers. Explanation in the text

2. Measuring (without the quarter-wave plate 7 and the polarizer-analyzer 8) the coordinate distribution $(m \times n)$ of the intensity of the image of biological tissue $I(m \times n)$, where $m \times n$ is the set of pixels of the digital camera 9.
3. The analyzer 8 is installed (in the absence of the quarter-wave plate 7), and the transmission axis is oriented at angles $\Theta = 0°$, $\Theta = 90°$ for each of which the distributions of the intensity values $I^{(0)}(m \times n)$, $I^{(90)}(m \times n)$ are measured.
4. The distributions of the Stokes vector parameters are measured [see this chapter, paragraph 2.2.1, relation (2.5)] for each individual pixel $(m, n)$ of the CCD camera—the functional sequence of polarization filtering is shown in fragment 12.
5. The polarization maps of the microscopic image of the histological section of the biological tissue are calculated using the relations (2.2), (2.3)—fragment 13.
6. The coordinate distribution of the phase shifts $\delta(m \times m)$ is calculated using the relation

$$\delta(r_{ik}) = \mathrm{arc}tg\left[\frac{tg2\beta(r_{ik})}{tg\alpha(r_{ik})}\right]. \tag{2.6}$$

Here, $r_{ik}$ is the coordinate of the local pixel in the plane of the photosensitive unit of the digital camera.

7. The coordinate distribution of the values of the CDMP modulus $|V|(m \times n)$-fragment 14—is calculated according to the relation [2]

$$|V| = \frac{\frac{\left(I_x^{0,5}(r_1)I_x^{0,5}(r_2) - I_y^{0,5}(r_1)I_y^{0,5}(r_2)\right)^4 + 16I_x(r_1)I_x(r_2)I_x(r_1)I_x(r_2)}{I(r_1)I(r_2)} +}{\frac{8I_x^{0,5}(r_1)I_x^{0,5}(r_2)I_x^{0,5}(r_1)I_x^{0,5}(r_2)\left(I_x^{0,5}(r_1)I_x^{0,5}(r_2) - I_y^{0,5}(r_1)I_y^{0,5}(r_2)\right)^2 \cos\delta_{12}}{I(r_1)I(r_2)}}; \tag{2.7}$$

where $I_x(r_1; r_2)$; $I_y(r_1; r_2)$ are the values of the intensity of the microscopic image of the histological section of biological tissue at points with coordinates

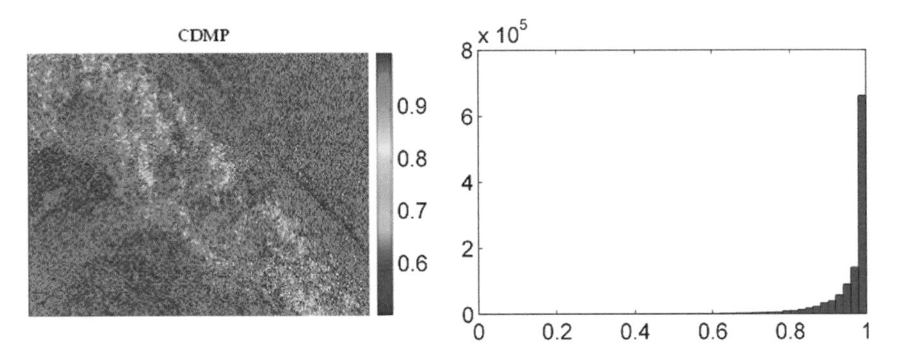


**Fig. 2.6** Coordinate distribution of values of the CDMP modulus of a microscopic image of a histological section of the rectal wall and a histogram of its random values

$r_1; r_2$ for the rotation of the transmission axis of the analyzer 9 relative to the plane of incidence at angles of 0° and 90°, respectively.

Figure 2.6 shows the coordinate distribution of the values of the CDMP modulus of the microscopic image of a histological section of the rectal wall and the distribution histogram.

Thus, the main result of the CDMP system mapping of microscopic images of histological sections of biological tissues is that it provides a new set of actions in relation to the correlation comparison of polarization parameters at different points of the optical field in the image plane. This leads to the expansion of the functionality capabilities of diagnostics of optical anisotropy of the biological layers and improves the information content by implementing the azimuthally invariant polarization-correlation process for measuring the coordinate distributions of the values of the CDMP modulus.

## 2.4 CDMA Mapping System

Figure 2.7 presents the optical scheme of the CDMA—mapping of optically anisotropic networks of histological sections of biological tissues.

To determine the values of the CDMA modulus, we can use the results of studies [8, 15]

$$W(r_1, r_2) = \frac{\{d_{11}(r_1, r_2) + d_{12}(r_1, r_2) + d_{21}(r_1, r_2) + d_{22}(r_1, r_2)\}^2}{I(r_1)I(r_2)} \tag{2.8}$$

Here, $d_{ik}(r_1, r_2)$ are the generalized matrix elements of the following for

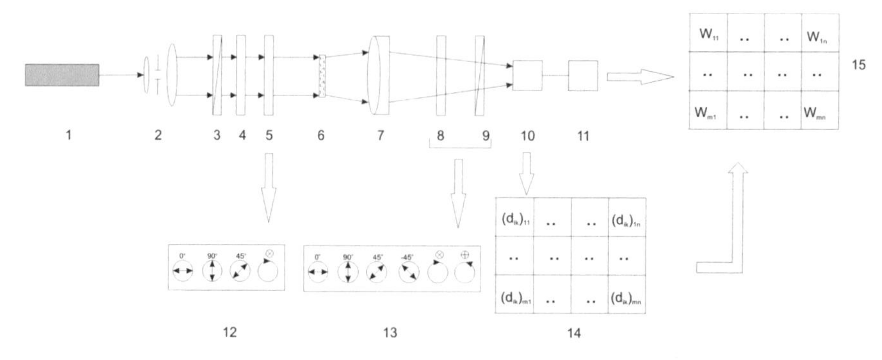


Fig. 2.7 Optical design of CDMA—mapping of biological layers. Explanation in the text

$$\begin{cases} d_{11}(r_1, r_2) = d_{11}(r_1)d_{11}(r_2); \\ d_{12}(r_1, r_2) = d_{12}(r_1)d_{12}(r_2); \\ d_{21}(r_1, r_2) = d_{21}(r_1)d_{21}(r_2); \\ d_{22}(r_1, r_2) = d_{22}(r_1)d_{22}(r_2), \end{cases} \tag{2.9}$$

where $d_{ik}$ are the complex elements of the Jones matrix $\{D\}$.

The experimental measurement of the real component $R_{ik}$ of complex elements $d_{ik}$ is based on the classical approach proposed in [13].

Values $R_{ik}$ are measured as follows:

- Irradiate sample 6 with a linearly polarized laser light beam 1 with azimuth $\alpha_0 = 0°$—the functional sequence of changing polarization states provides a filter, which consists of quarter-wave plates 3, 5 and polarizer 4, illustrates fragment 12;
- Rotate the transmission axis of the polarizer-analyzer 9 (in the absence of a quarter-wave plate 8) the angles $\Theta = 0°$, $\Theta = 90°$ and measure the intensity of the transmitted radiation $I_0^0$; $I_{90}^0$—the functional sequence of changes in polarization states, provides a polarization analyzer, which consists of a quarter-wave plate 8 and polarizer 9, illustrates fragment 13;
- Irradiate the sample 6 with a linearly polarized light beam with azimuth $\alpha_0 = 90°$;
- Rotate the axis of transmission of the polarizer the corners $\Theta = 0°$, $\Theta = 90°$ and measure the intensity of the transmitted radiation $I_0^{90}$; $I_{90}^{90}$;
- Calculate for each pixel of the digital camera 10 the real components $R_{ik}$ of the elements of the Jones matrix—fragment 14

$$\left\{ R_{11} = \sqrt{I_0^0};\ R_{12} = \sqrt{I_{90}^0};\ R_{21} = \sqrt{I_0^{90}};\ R_{22} = \sqrt{I_{90}^{90}}. \right. \tag{2.10}$$

- Calculate the value of the CDMA modulus—fragment 15 for each pixel of a digital camera using the relations (2.8)–(2.10).

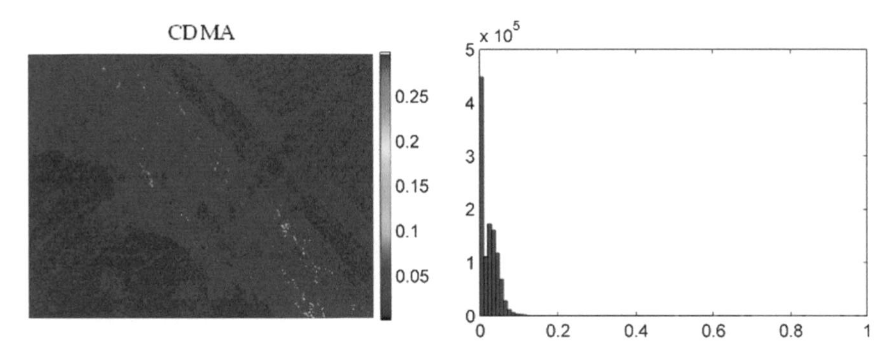


**Fig. 2.8** Coordinate distribution of the values of the CDMA modulus of the polycrystalline structure of a histological section of the rectal wall and a histogram of the distribution of its random values

Figure 2.8 shows the coordinate distribution of the values of the CDMA modulus of the polycrystalline structure of the histological section of the rectal wall and the distribution histogram.

Hence, the main result of the CDMA system mapping of polycrystalline networks of histological sections of biological tissues is that it provides a new set of actions in relation to the correlation comparison of the values of the real component of the elements of the Jones matrix at different points on the plane of the object. This leads to the expansion of the functional capabilities of the optical anisotropy of biological layers and to the improvement of information content by implementing an azimuthally invariant polarization-correlation process for measuring coordinate distributions of CDMA modulus values.

## 2.5 Wavelet Analysis Scheme

Figure 2.9 shows the results of a wavelet analysis (central column) of the two-dimensional azimuth and polarization elliptic distributions (left column) of a microscopic image of a histological section of a prostate adenoma sample. The right-hand column shows linear sections of the two-dimensional distribution of wavelet-coefficients.

The main result of the wavelet analysis of the distributions of polarization (azimuth, ellipticity), Mueller-matrix (azimuthally invariant matrix elements and their combinations), polarization-correlation (CDMP, CDMA) parameters is that it provides a new set of actions for assessment of the structure of polycrystalline networks of histological sections of biological tissues at different scales of geometric dimensions. This leads to the expansion of the functional capabilities of the optical anisotropy of biological layers and improves information content.

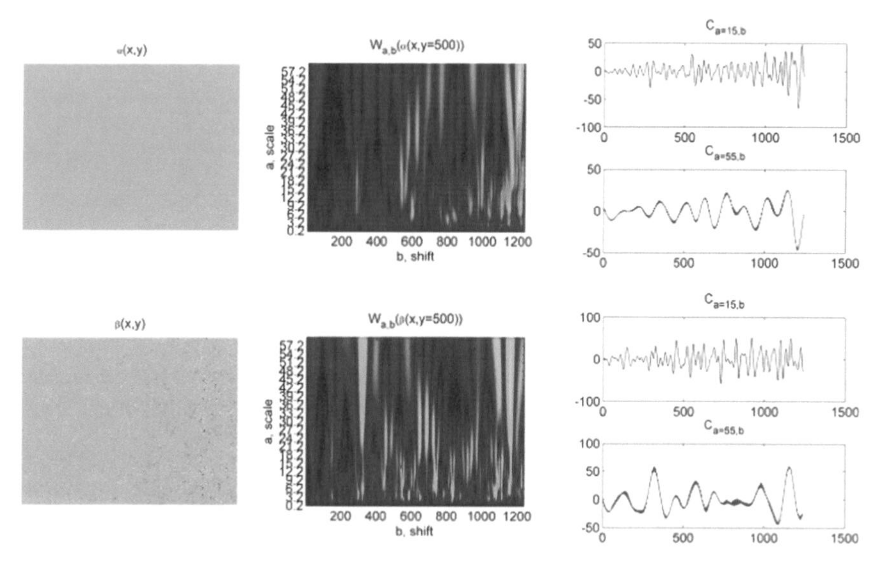


**Fig. 2.9** An example of a wavelet analysis of polarization maps of azimuth (upper line) and ellipticity (lower line) of a microscopic image of a histological section of prostate adenoma

## 2.6 Fourier Analysis Scheme

Figure 2.10 shows the optical scheme of the azimuthally invariant Fourier–Stokes polarimetry of microscopic images of histological sections of biological tissues.

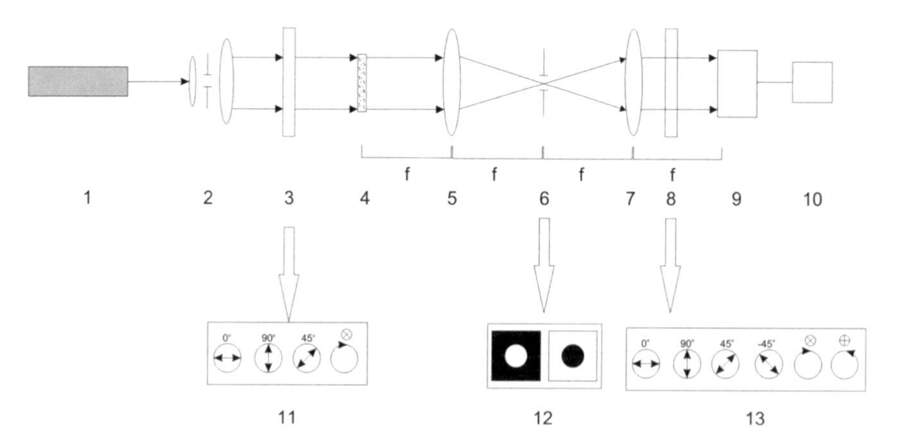


**Fig. 2.10** Optical scheme of an azimuthally invariant Stokes polarimeter using spatial-frequency filtering, where 1 is a He–Ne laser; 2—collimator; 3—polarization filter with functional algorithm 11; 4—object of study; 5, 7—microlenses; 6, 12—low-frequency-pass and high-frequency-pass filters, 8—polarization filter with a functional algorithm 13; 9—CCD camera; 10—personal computer

The illumination of the test sample 4 was carried out by a parallel (Ø = 104 μm) He–Ne laser beam 1 ($\lambda = 0.6328\,\mu$m). The illuminator 3 consists of two quarter-wave plates and a polarizer and ensures the formation of the right circularly polarized laser beam—fragment 11. Histological sections of biological tissues 4 were located in the focal plane of a polarizing microlens 5 (focal length $f = 30$ mm, magnification 4×, digital aperture $N.A. = 0.1$). In the back focal plane, there was a spatial-frequency (low-frequency or high-frequency) filter 6, 12. The polarizing microlens 7 (focal length $f = 30$ mm, 4× magnification, digital aperture $N.A. = 0.1$) was located at the focal length from the frequency plane of the lens 7 and, therefore, performed the inverse Fourier transform of the spatially frequency filtered laser radiation field. The coordinate intensity distribution of such a field was recorded in the plane of the photosensitive CCD camera 9, which was also at the focal distance from the microlens 7.

In Fig. 2.11, the results of spatial-frequency filtering of the coordinate distributions of the values of the fourth parameter of the Stokes vector of a microscopic image of a histological section of the rectum wall are presented.

The main result of the Fourier analysis of the distributions of polarization (azimuth, ellipticity), Mueller-matrix (azimuthally invariant matrix elements and their combinations), and polarization-correlation (CDMP, CDMA) parameters is that it provides a new set of actions, direct experimental assessment of the structure of polycrystalline networks of histological sections of biological tissues at different scales of geometric dimensions. This leads to the expansion of the functionality of the diagnosis of optical anisotropy of biological layers and improves information content.

## 2.7 Characterization of Research Objects

Optically thin (geometric thickness $d = 15$–$35\,\mu$m, attenuation coefficient $\tau < 0.1$) histological sections of human biological tissues prepared according to the standard protocol on a freezing microtome were used as research objects.

From the optical point of view, such samples are layers with the presence of an optically anisotropic component in the form of birefringent fibrillar networks. The specified choice of objects allows us to analyze the possibilities of polarization Stokes correlometry from the point of view of fundamental and applied applications.

*Fundamental application*—during the propagation of laser radiation through such samples, a single interaction is realized with spatially oriented fibrils and optically anisotropic networks created by them. As a result of this interaction, the phenomenon of modulation of the azimuth and elliptic polarization of partial laser waves arises. A manifestation of this interaction is the formation of a polarization-inhomogeneous image of such a biological layer. A characteristic difference of this process is the presence of various mechanisms of optical anisotropy, which is realized at different scales of fibrillar structures. Therefore, to determine the diagnostically relevant unique relationship between the polarization states at the points of

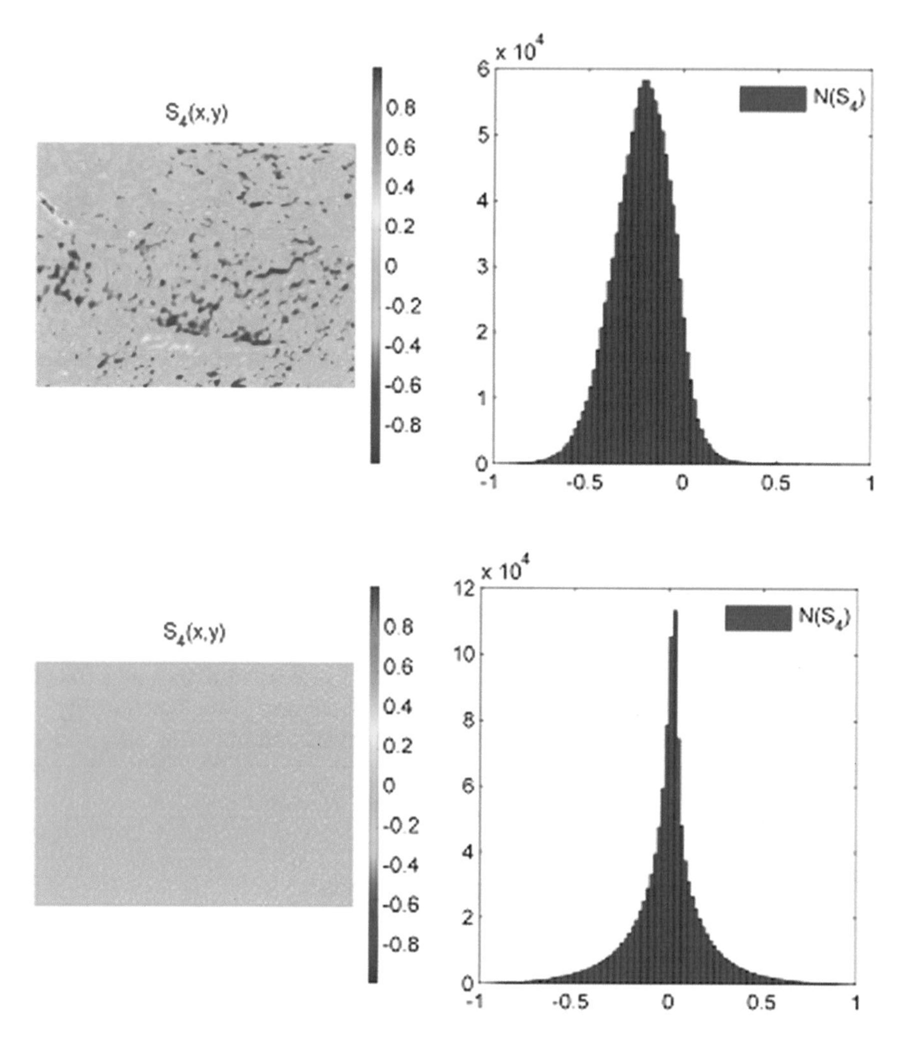


**Fig. 2.11** Low-frequency (upper line) and high-frequency (lower line) components of the coordinate distributions of the values of the fourth parameter of the Stokes vector of a microscopic image of a histological section of the rectal wall and a histogram of the distribution of random values

such an image and the optical anisotropy parameters (birefringence is the direction of the optical axis, phase shift between the orthogonal polarization components; optical activity is the angle of rotation of the polarization plane), the correlation approach is extremely important. This approach in our work is a combination of interconnected methods of "mathematical (wavelet analysis)" and "physical (spatial-frequency filtering)" convolution and polarization-correlation CDMP and CDMA mapping.

*Applied*—histological sections of biological tissues are a classic object of clinical histological microscopic studies, accepted currently as a "gold" standard. Based on this, Stokes-correlometric studies are additional objective methods, and in what follows, we will call them "polarization-correlation biopsy".

As a sample of human biological tissues in the "fundamental approximation," we used histological sections with an ordered (skeletal muscle—Fig. 2.12) and a disordered (brain—Fig. 2.13) structure of birefringent fibrillar networks.

In the "applied approximation," we examined histological sections of the myocardium of subjects who had died due to mechanical asphyxia (Fig. 2.14) and heart attack (Fig. 2.15).

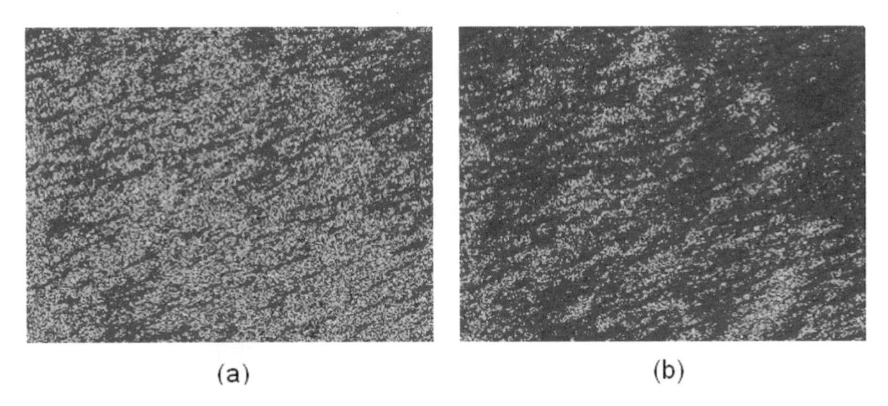


**Fig. 2.12** Microscopic images of a histological section of skeletal muscle tissue obtained in coaxial (0°–0°) (**a**) and crossed (0°–90°) (**b**) polarizer-analyzer

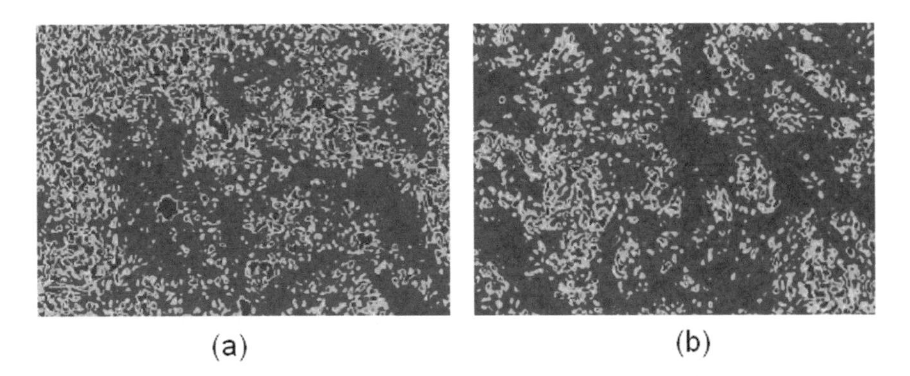


**Fig. 2.13** Microscopic images of a histological section of brain tissue obtained in coaxial (0°–0°) (**a**) and crossed (0°–90°) (**b**) polarizer-analyzer

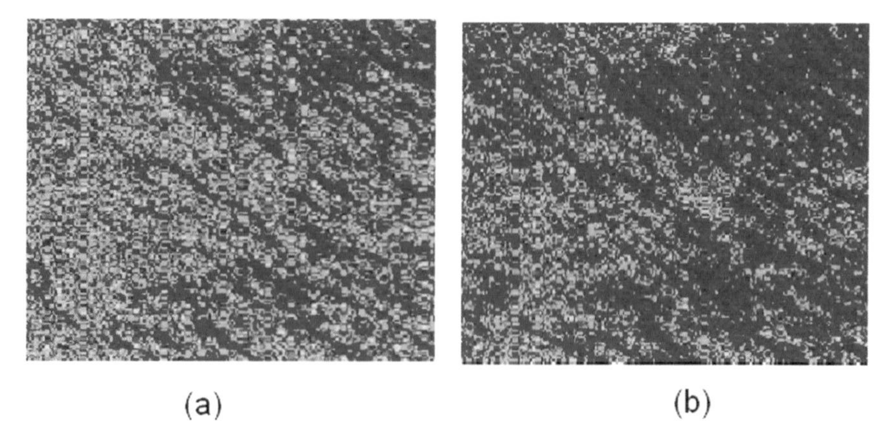


**Fig. 2.14** Microscopic images of a histological section of the deceased's myocardium due to mechanical asphyxia obtained in coaxial (0°–0°) (**a**) and crossed (0°–90°) (**b**) polarizer-analyzer

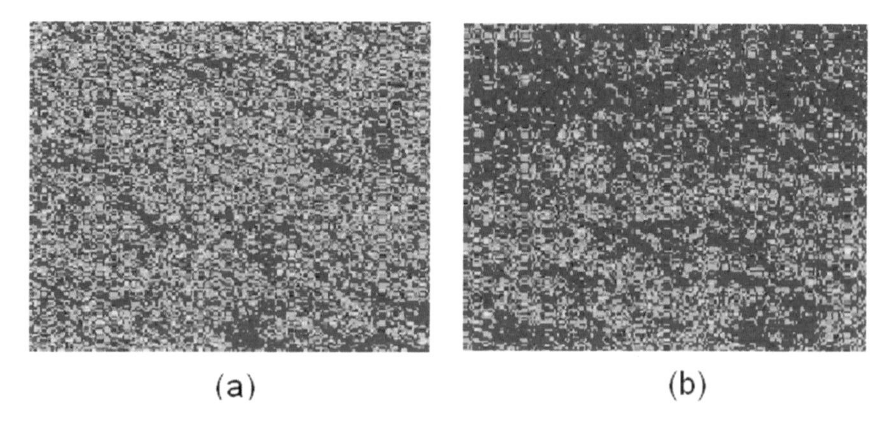


**Fig. 2.15** Microscopic images of a histological section of a myocardium that died due to a heart attack, obtained in coaxial (0°–0°) (**a**) and crossed (0°–90°) (**b**) polarizer-analyzer

# References


1. Y.O.Ushenko, Y.Y. Tomka, O.I. Telenga, I.Z. Misevitch, V.V. Istratiy, Complex degree of mutual anisotropy of biological liquid crystals nets. Opt. Eng. **50,** 039001 (2011)
2. Y.A. Ushenko, O.I. Telenga, A.P. Peresunko, O.K. Numan, New parameter for describing and analyzing the optical-anisotropic properties of biological tissues. J. Innov. Opt. Health Sci. **4**(4), 463–475 (2011)
3. A.V. Dubolazov, O.Y. Telenha, V.A. Ushenko, M.I. Sydor, Characteristic values of Mueller-matrixes images of biological liquid crystals net for diagnostics of human tissues anisotropy. Proc. SPIE. **8338**, 83380Z (2011)
4. O.Y. Novakovska, Polarization correlometry of characteristic states of Muller-matrix images of phase-inhomogeneous biological layers. Semicond. Phys. Quantum Electron. Optoelectron. **15**(3), 230–237 (2012)

5. A.G. Ushenko, Y.A. Ushenko, Y.Y. Tomka, O.V. Dubolazov, O.Y. Telenga, V.I. Istratiy, A.O. Karachevtsev, The interconnection between the coordinate distribution of Muller-matrixes images characteristics values of biological liquid crystals net and the pathological changes of human tissues : 12–16 July 2010, 9th Euro-American Workshop on Information Optics. Helsinki, Finland (2010)
6. Y.A. Ushenko, O.V. Dubolazov, OYu. Telenga, A.P. Angelsky, A.O. Karachevtcev, V. Balanetska, Complex degree of mutual anisotropy of biological liquid crystals net. Proc. SPIE. **8087**, 80872Q (2011)
7. O.V. Angelsky, Yu.A. Ushenko, V.O. Balanetska, The degree of mutual anisotropy of biological liquids polycrystalline nets as a parameter in diagnostics and differentiations of hominal inflammatory processes. Proc. SPIE. **8338**, 83380S (2011)
8. Yu.A. Ushenko, A.V. Dubolazov, A.O. Karachevtcev, N.I. Zabolotna, Complex degree of mutual anisotropy in diagnostics of biological tissues physiological changes. Proc. SPIE. **8134**, 81340O (2011)
9. Y.Y. Tomka, Wavelet analysis of biological tissue's Mueller-matrix images. Proc. SPIE. **7008**, 700823 (2008)
10. O.V. Dubolazov, Y.O. Ushenko, Y.Y. Tomka, O.G. Pridiy, A.V. Motrich, I.Z. Misevitch, V.V. Istratiy, Wavelet analysis for Mueller matrix images of biological crystal networks. Semiconductor Phys. Quantum Electron. Optoelectron. **12**(4), 391–398 (2009)
11. A.O. Karachevtsev, Fourier Stokes-polarimetry of biological layers polycrystalline networks. Semicond. Phys. Quantum Electron. Optoelectron. **15**(3), 252–268 (2012)
12. A. Doronin, C. Macdonald, I. Meglinski, Propagation of coherent polarized light in turbid highly scattering medium. J. Biomed. Opt. **19**(2), 025005 (2014)
13. A. Doronin, A. Radosevich, V. Backman, I. Meglinski, Two electric field Monte Carlo models of coherent backscattering of polarized light. J. Opt. Soc. America A **31**(11), 2394 (2014)
14. A. Ushenko, V. Pishak, in *Laser Polarimetry of Biological Tissue: Principles and Applications*, ed. by V. Tuchin. Handbook of Coherent-Domain Optical Methods: Biomedical Diagnostics (Environmental and Material Science, 2004) , pp. 93–138
15. S.B. Yermolenko, C.Y. Zenkova, A.P. Angelskiy, Polarization manifestations of correlation (intrinsic coherence) of optical fields. Appl. Opt. **47**(32) (2008)

# Chapter 3
# Scale-Selective and Spatial-Frequency Correlometry of Polarization-Inhomogeneous Field

## 3.1 Wavelet Analysis of Azimuthally Invariant Distributions of Polarization Parameters of Microscopic Images of Biological Tissues

### *3.1.1 Wavelet Analysis of Azimuthally Invariant Polarization Maps of Spatially Ordered Optically Anisotropic Networks of Biological Tissues*

Figures 3.1, 3.2, and Table 3.1 present the results of a wavelet analysis of polarization maps of microscopic images of a histological section of the skeletal muscle.

Analysis of the data revealed a different "structure" of the distribution of the values of the wavelet-coefficients $W_{a,b}\{\alpha(x, y); \beta(x, y)\}$ and their different-scale anharmonic distributions $C_{a,b}$. This result can be associated with the structure of the optically anisotropic spatially ordered fibrillar network of skeletal muscle.

At small scales, formed by optically active chains of molecules, high-frequency modulation of the magnitude of circular birefringence predominates, which is manifested in the formation of the corresponding anharmonic modulation of the amplitude of the wavelet-coefficients $W_{a=\min,b}$ of the polarization azimuth map at the points of the microscopic image. The reverse picture holds for the large scale of the MHAT function window which highlights the manifestations of linear birefringence—a polarization map of ellipticity. Quantitatively, the possibility of an objective assessment of the manifestations of birefringence at different scales of the dimensions of a spatially structured fibrillar network is illustrated in Table 3.1.

An analysis of the data revealed that all the statistical moments that characterize the distribution of the values of the wavelet-coefficients $W_{a,b}\{\alpha(x, y); \beta(x, y)\}$ are nonzero. It was revealed that the values of statistical moments of the third to fourth orders that characterize the distribution of the amplitudes of the wavelet-coefficients



I. Meglinski et al., *Shedding the Polarized Light on Biological Tissues*,
SpringerBriefs in Applied Sciences and Technology,
https://doi.org/10.1007/978-981-10-4047-4_3

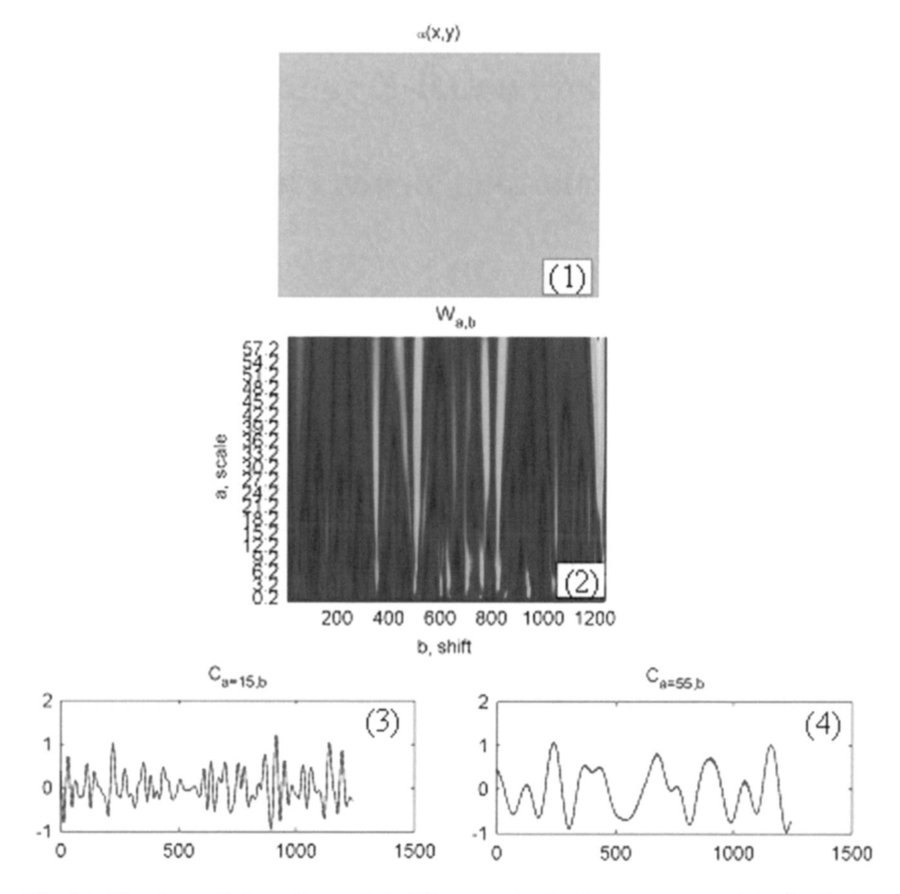


**Fig. 3.1** Wavelet-coefficients (2) and their different-scale (3), (4) cross sections of the distribution of polarization azimuth (1) of the microscopic image of a histological section of skeletal muscle

$W_{a,b}$ in the region of small ($W_{a,b}\{\alpha(x, y)\}$) and large ($W_{a,b}\{\beta(x, y)\}$) window scales of the wavelet function are most dynamically changed.

### 3.1.2 Wavelet Analysis of Azimuthally Invariant Polarization Maps of Spatially Disordered Optically Anisotropic Networks of Biological Tissues

Figures 3.3, 3.4 and Table 3.2 present the results of a wavelet analysis of polarization maps of microscopic images of a histological section of the brain.

A comparative analysis of the results of wavelet scanning data of two-dimensional distributions of polarization maps of a microscopic image of a histological section of the brain and the similar study of a birefringent fibrillar network of a histological

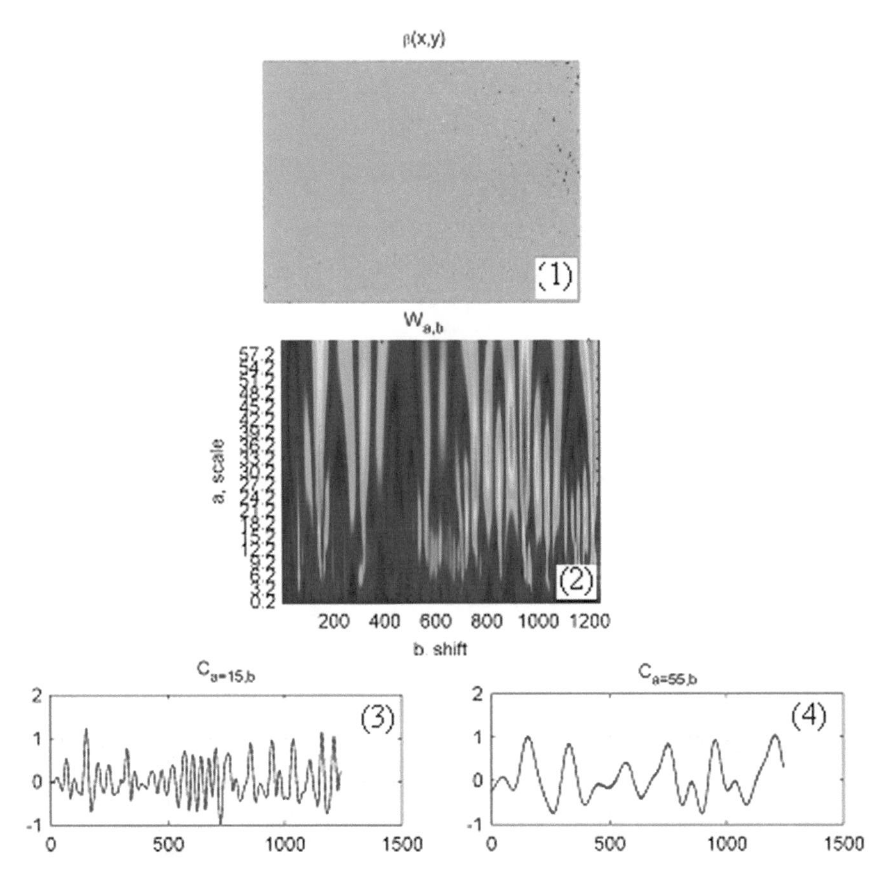


**Fig. 3.2** Wavelet-coefficients (2) and their different-scale (3), (4) cross sections of the distribution of polarization ellipticity (1) of the microscopic image of a histological section of skeletal muscle

**Table 3.1** Statistical moments $Q_{i=1;2;3;4}$ that characterize the distribution of the values of the wavelet-coefficients of the polarization maps of azimuth $\alpha(m \times n)$ and ellipticity $\beta(m \times n)$ at the points of the microscopic image of the histological section of skeletal muscle

| Parameters | $\alpha(m \times n)$ | $\beta(m \times n)$ | $\alpha(m \times n)$ | $\beta(m \times n)$ | $\alpha(m \times n)$ | $\beta(m \times n)$ |
|---|---|---|---|---|---|---|
| $a$ | 15 | | 56 | | 80 | |
| $Q_{i=1}$ | 0.01 | 0.07 | 0.06 | 0.08 | 0.08 | 0.11 |
| $Q_{i=2}$ | 0.18 | 0.14 | 0.15 | 0.12 | 0.11 | 0.08 |
| $Q_{i=3}$ | 0.29 | 0.46 | 0.72 | 0.98 | 1.12 | 1.38 |
| $Q_{i=4}$ | 0.46 | 0.52 | 0.37 | 0.41 | 0.25 | 0.21 |

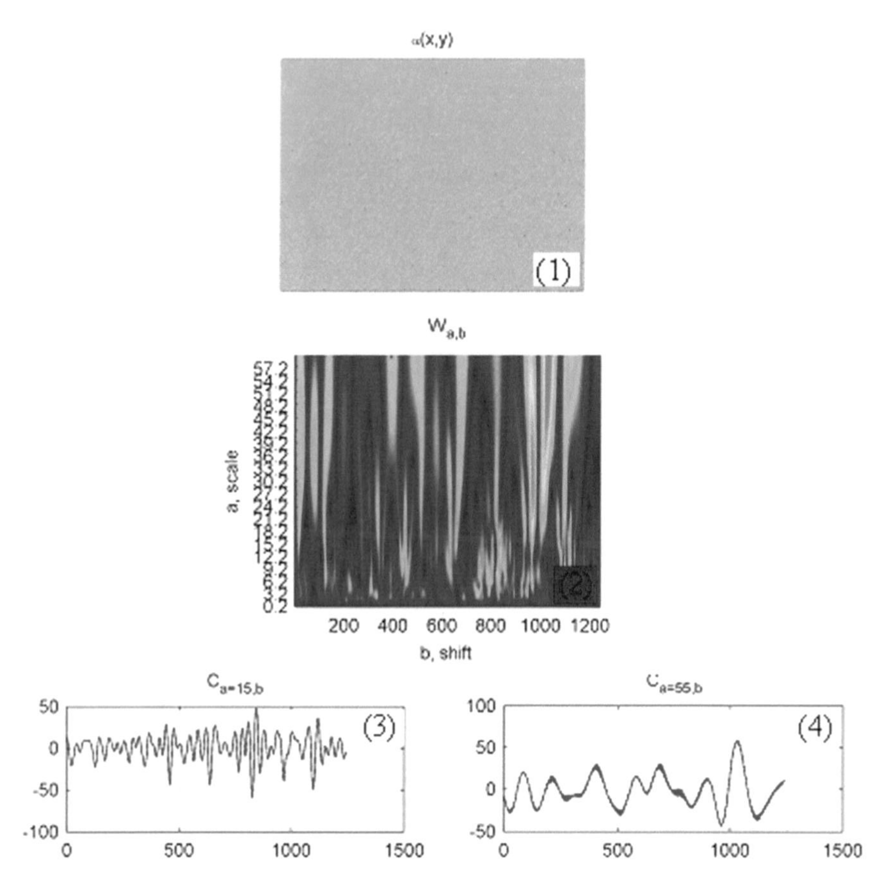


**Fig. 3.3** Wavelet-coefficients (2) and their different-scale (3), (4) cross sections of the distribution of polarization azimuth (1) of a microscopic image of a histological section of the brain

section of the skeletal muscle revealed the formation of anharmonic distributions $C_{a,b}$ at all scales $a$ of the MHAT functions window. At the same time, the specificity of the optical-geometric structure of the disordered network of optical-anisotropic protein fibers of brain tissue is manifested in a more "smooth" change in the amplitudes of the wavelet-coefficients $W_{a,b}$ at different window scales $a$ of the wavelet function $\Omega_{a,b}$, which scans azimuthally invariant polarization maps $\{\alpha(x, y); \Delta\beta(x, y)\}$. In accordance with this, other values of statistical moments of the first to fourth orders are formed that characterize such distributions—Table 3.2.

From the data in Table 3.2, it can be seen that the differences between the values of all statistical moments of the first to fourth orders, which characterize different-scale sections $C_{a=15,b}$ and $C_{a55,b}$ of the two-dimensional distributions of wavelet-coefficients $W_{a,b}\{\alpha(x, y); \beta(x, y)\}$, are two to four times less than those shown in Table 3.1.

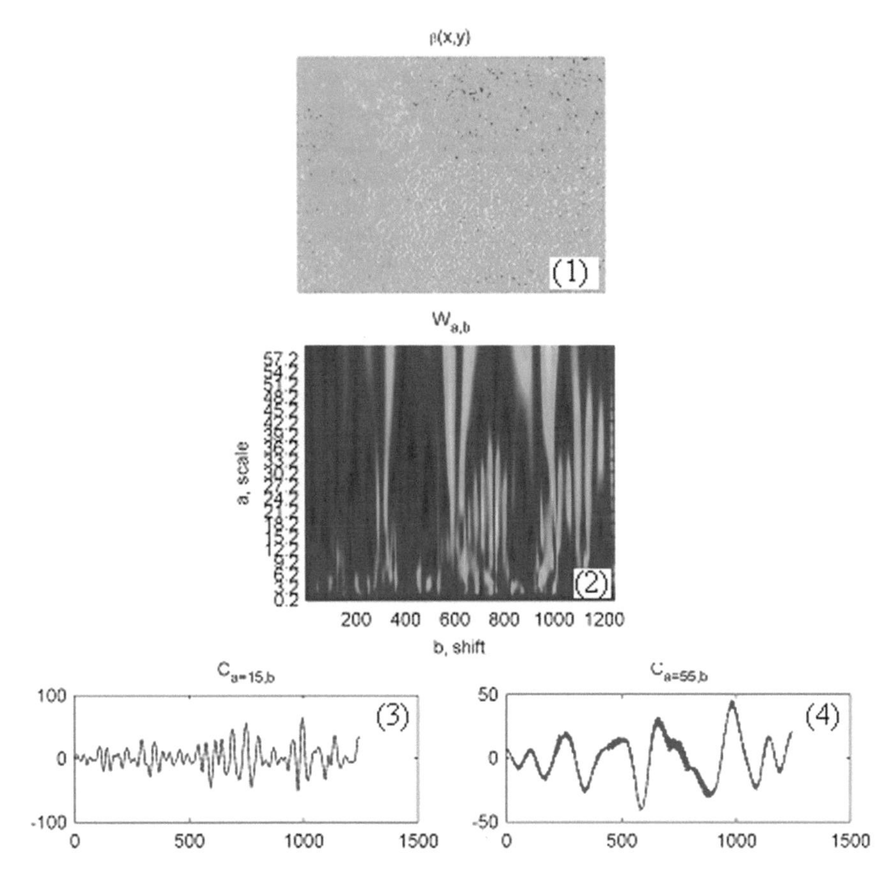


**Fig. 3.4** Wavelet coefficients (2) and their different-scale (3), (4) cross sections of the distribution of polarization ellipticity (1) of a microscopic image of a histological section of the brain

**Table 3.2** Statistical moments $Q_{i=1;2;3;4}$ that characterize the distribution of the values of the wavelet-coefficients of the polarization maps of azimuth $\alpha(m \times n)$ and ellipticity $\beta(m \times n)$ at points of a microscopic image of a histological section of the brain

| Parameters | $\alpha(m \times n)$ | $\beta(m \times n)$ | $\alpha(m \times n)$ | $\beta(m \times n)$ | $\alpha(m \times n)$ | $\beta(m \times n)$ |
|---|---|---|---|---|---|---|
| $a$ | 5 | | 25 | | 80 | |
| $Q_{i=1}$ | 0.01 | 0.07 | 0.06 | 0.08 | 0.08 | 0.11 |
| $Q_{i=2}$ | 0.14 | 0.11 | 0.11 | 0.08 | 0.07 | 0.05 |
| $Q_{i=3}$ | 0.19 | 0.34 | 0.47 | 0.79 | 0.81 | 0.92 |
| $Q_{i=4}$ | 0.33 | 0.41 | 0.22 | 0.31 | 0.15 | 0.14 |

Thus, we have detected sensitivity at different scales of wavelet scanning of two-dimensional distributions of azimuth and ellipticity of polarization of microscopic images of polycrystalline networks of biological tissues of various types.

The information obtained was the basis for the development of a method of polarization-correlation microscopy of weak changes in optical anisotropy due to necrotic changes in the myocardium of subjects deceased due to various reasons—specifically, mechanical asphyxiation and heart attack.

### 3.1.3 Diagnostic Application of Wavelet Analysis of Azimuthally Invariant Polarization Maps

The results of the wavelet analysis of azimuthally invariant polarization maps of microscopic images of myocardium histological sections deceased due to mechanical asphyxiation and heart attack are shown in Figs. 3.5 and 3.6.

A comparative analysis of the results of the study of wavelet-coefficients $W_{a,b}(\alpha)$ revealed the greatest differences between them at the level of small scales $a_{\min}$ of the MHAT function. This fact is indicated by the large modulation depth of the "small-scale" dependences $C_{a=5,15,b}$ that describe the circular birefringence of myosin molecules.

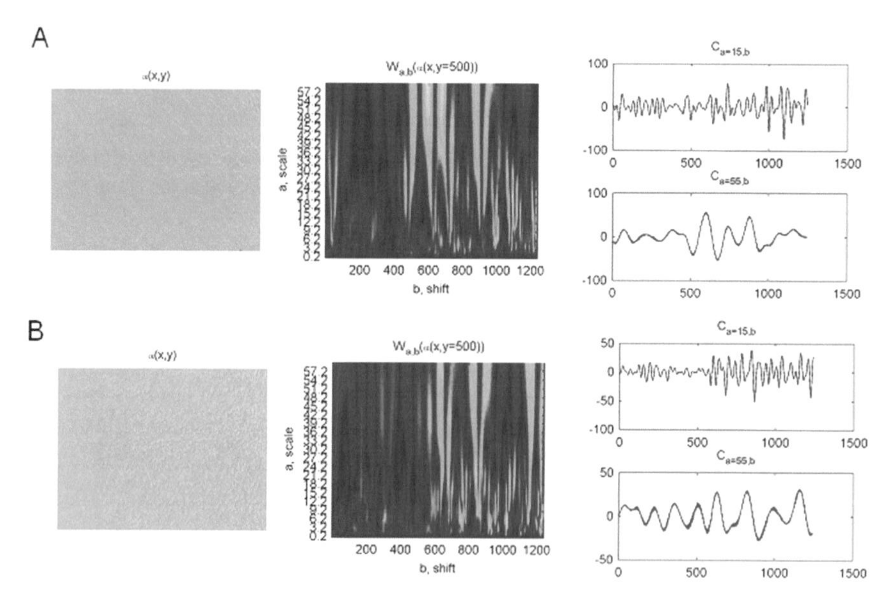


**Fig. 3.5** Wavelet-coefficients and their different-scale cross sections of the distribution of the polarization azimuth of microscopic images of myocardium histological sections deceased due to mechanical asphyxiation (fragment A) and heart attack (fragment B)

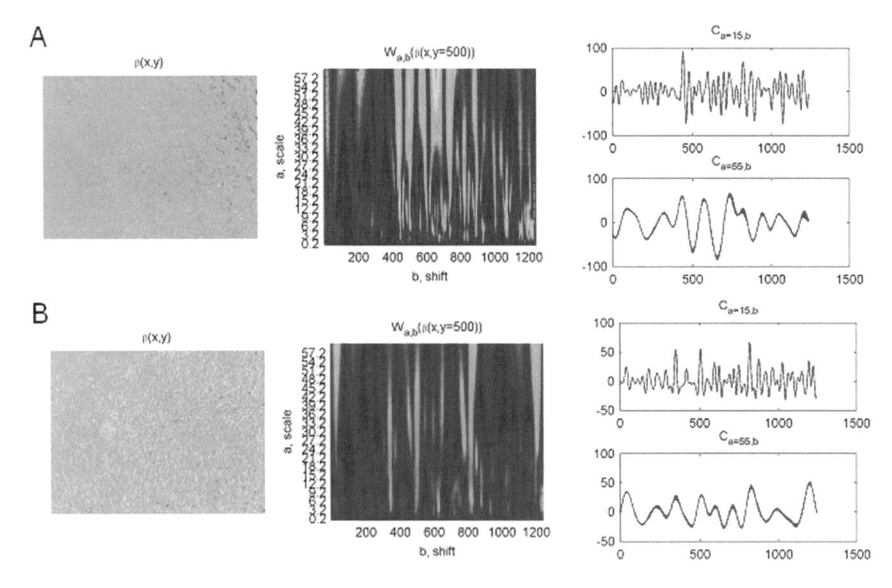


**Fig. 3.6** Wavelet-coefficients and their different-scale cross sections of the distribution of polarization ellipticity of microscopic images of myocardium histological sections deceased due to mechanical asphyxiation (fragment A) and heart attack (fragment B)

For the polarization ellipticity map, such differences occur for "large-scale" dependences $C_{a=80,b}$ that describe the linear birefringence of a network of spatially ordered myosin fibrils.

The revealed patterns can be associated with the fact that necrotic changes in the myocardium polycrystalline networks of the deceased due to mechanical asphyxiation occur not at the morphological (large-scale), but at the concentration (small-scale) level. Quantitatively, this process manifests itself in the growth of higher-order statistical moments $Q_{i=3;4}\big(C_{a=5,b}(\alpha)\big) \uparrow$ characterizing the distribution of the values of the small-scale cross section $C_{a=5,b}$ of a two-dimensional wavelet map $W_{a,b}\{\alpha(x, y)\}$—Table 3.3.

For the samples deceased due to a heart attack, myosin networks are destroyed, which is manifested in a change in the modulation of the magnitude of the amplitude of the map of "large-scale" cross sections of the wavelet-coefficient $W_{a,b}\{\beta(x, y)\}$—Table 3.4.

The following differences (highlighted in italic) between the statistical parameters that characterize the distribution of values of different-scale cross sections of the values of wavelet maps $W_{a,b}\{\alpha(x, y); \beta(x, y)\}$ of polycrystalline networks of samples of the myocardium histological sections of both types

$\alpha(m \times n) \Leftrightarrow a_{\min} \rightarrow \Delta M_3 = 1.39; \Delta M_4 = 1.25;$
$\beta(m \times n) \Leftrightarrow a_{\max} \rightarrow \Delta M_3 = 1.36; \Delta M_4 = 1.22$
are established.

In this case, a high level of balanced accuracy ($Ac = 73$–$76\%$) of the proposed method for the differentiation of precancerous conditions is achieved. According to

**Table 3.3** Statistical moments $Q_{i=1;2;3;4}$ that characterize the distribution of the values of the wavelet-coefficients of the polarization maps of azimuth at the points of microscopic images of the myocardium histological sections deceased due to mechanical asphyxiation ($\alpha(m \times n)$) and heart attack $\alpha^*(m \times n)$ and the balanced accuracy of the method

| Parameters | $\alpha(m \times n)$ | $\alpha^*(m \times n)$ | $\alpha(m \times n)$ | $\alpha^*(m \times n)$ | $Ac(\%)$ | $Ac(\%)$ |
|---|---|---|---|---|---|---|
| $a$ | 5 | | 80 | | 5 | 80 |
| $Q_{i=1}$ | 0.012 ± 0.0018 | 0.01 ± 0.0013 | 0.0095 ± 0.0001 | 0.008 ± 0.0001 | 51 | 52 |
| $Q_{i=2}$ | 0.19 ± 0.023 | 0.16 ± 0.022 | 0.11 ± 0.016 | 0.097 ± 0.011 | 54 | 53 |
| $Q_{i=3}$ | *0.39 ± 0.044* | *0.23 ± 0.029* | 1.01 ± 0.14 | 0.91 ± 0.12 | 73 | 67 |
| $Q_{i=4}$ | *0.55 ± 0.063* | *0.36 ± 0.047* | 0.21 ± 0.032 | 0.15 ± 0.022 | 76 | 63 |

**Table 3.4** Statistical moments $Q_{i=1;2;3;4}$ that characterize the distribution of the values of the wavelet-coefficients of the polarization maps of azimuth at the points of microscopic images of the myocardium histological sections deceased due to mechanical asphyxiation ($\beta(m \times n)$) and heart attack ($\beta^*(m \times n)$) and the balanced accuracy of the method

| Parameters | $\beta(m \times n)$ | $\beta^*(m \times n)$ | $\beta(m \times n)$ | $\beta^*(m \times n)$ | $Ac(\%)$ | $Ac(\%)$ |
|---|---|---|---|---|---|---|
| $a$ | 5 | | 80 | | 5 | 80 |
| $Q_{i=1}$ | 0.07 ± 0.09 | 0.06 ± 0.008 | 0.11 ± 0.015 | 0.105 ± 0.014 | 52 | 53 |
| $Q_{i=2}$ | 0.14 ± 0.019 | 0.12 ± 0.017 | 0.08 ± 0.001 | 0.065 ± 0.008 | 56 | 54 |
| $Q_{i=3}$ | 0.46 ± 0.056 | 0.38 ± 0.045 | *1.38 ± 0.17* | *1.03 ± 0.13* | 65 | 74 |
| $Q_{i=4}$ | 0.52 ± 0.063 | 0.44 ± 0.052 | *0.21 ± 0.032* | *0.14 ± 0.019* | 66 | 76 |

the criteria of evidence-based medicine, such accuracy corresponds to a good-quality diagnostic test.

Note that the balanced accuracy of direct azimuthally invariant polarization mapping of microscopic images of histological sections of the myocardium with various causes of death is significantly lower—$Ac \sim 60\%$.

## 3.2 Wavelet Analysis of Azimuthally Invariant Mueller-Matrix Images of Biological Tissues

### 3.2.1 Wavelet Analysis of Azimuthally Invariant Mueller-Matrix Images of Spatially Ordered Optically Anisotropic Networks of Biological Tissues

Figures 3.7, 3.8, and Table 3.5 present the results of a wavelet analysis of azimuthally invariant Mueller-matrix images of spatially ordered optically anisotropic networks of a histological section of the skeletal muscle.

Analysis of the data revealed differences in the modulation depth of different-scale cross sections $C_{a=15,b}$ and $C_{a=55,b}$ wavelet-coefficients $W_{a,b}\{f_{44}(x, y); \Delta f(x, y)\}$. As in the case of the wavelet analysis of polarization maps of the histological section

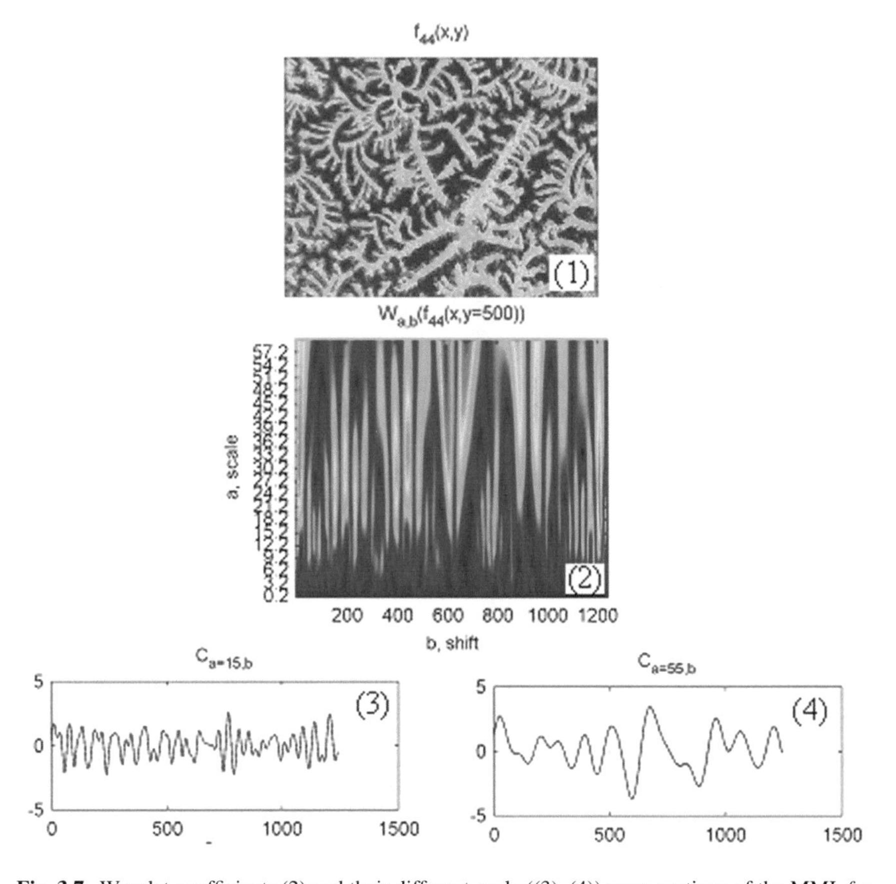


**Fig. 3.7** Wavelet-coefficients (2) and their different-scale ((3), (4)) cross sections of the MMI $f_{44}$ distribution (1) of the histological section of the skeletal muscle

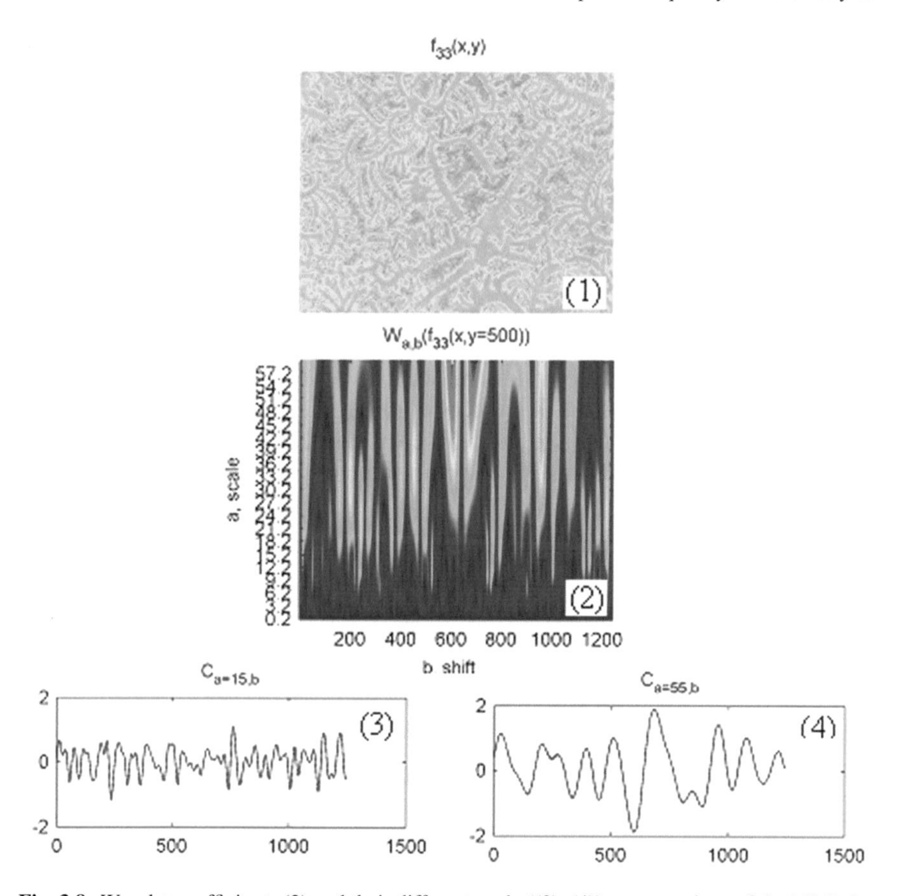


**Fig. 3.8** Wavelet-coefficients (2) and their different-scale ((3), (4)) cross sections of the MMI $f_{33}$ distribution (1) of the histological section of the skeletal muscle

**Table 3.5** Statistical moments $Q_{i=1;2;3;4}$ that characterize the distribution of the values of the wavelet coefficients of the MMI $f_{44}(m \times n)$ and ellipticity $\Delta f(m \times n)$ at the points of the microscopic image of the histological section of the skeletal muscle

| Parameters | $f_{44}(m \times n)$ | $\Delta f(m \times n)$ | $f_{44}(m \times n)$ | $\Delta f(m \times n)$ | $f_{44}(m \times n)$ | $\Delta f(m \times n)$ |
|---|---|---|---|---|---|---|
| $a$ | 5 | | 25 | | 80 | |
| $Q_{i=1}$ | 0.04 | 0.025 | 0.02 | 0.045 | 0.015 | 0.085 |
| $Q_{i=2}$ | 0.42 | 0.28 | 0.36 | 0.23 | 0.22 | 0.15 |
| $Q_{i=3}$ | 0.31 | 0.15 | 0.39 | 0.11 | 0.54 | 0.06 |
| $Q_{i=4}$ | 0.17 | 1.09 | 0.79 | 0.87 | 1.19 | 0.62 |

of the skeletal muscle, the greatest modulation depth for linear dependencies $C_{a,b}$ characterizing the distribution $\{\Delta f(x, y)\}$ is established, which is directly interconnected with the circular birefringence of myosin chains. At large scales of the MHAT-function window, the modulation of the distribution cross sections of another MMI $\{f_{44}(x, y)\}$, which characterizes the linear birefringence of the skeletal muscle fibrillar network, increases. Table 3.5 shows the values of statistical moments of the first to fourth orders characterizing the distribution of the values $C_{a=15,b}$ and $C_{a55,b}$ calculated wavelet-coefficients of the Mueller-matrix images $(f_{44}, \Delta f)$ of the histological section of the skeletal muscle.

Analysis of the obtained data revealed that all statistical moments that characterize the distribution of values of different-scale linear cross sections of two-dimensional wavelet-coefficients $W_{a,b}\{f_{44}(x, y); \Delta f(x, y)\}$ are nonzero.

It was revealed, as in the case of polarization mapping, that the greatest dynamics of changes in the statistical moments of the third and fourth orders characterizing the distribution of the amplitudes of the wavelet-coefficients $W_{a,b}$in the region of small $(W_{a,b}\{\Delta f(x, y)\})$ and large $(W_{a,b}\{f_{44}(x, y)\})$ scales of the window of wavelet function.

### *3.2.2 Wavelet Analysis of Azimuthally Invariant Mueller-Matrix images of Spatially Disordered Optically Anisotropic Networks of Biological Tissues*

Figures 3.9, 3.10, and Table 3.6 present the results of a wavelet analysis of polarization maps of microscopic images of a histological section of the brain.

A comparison of the results of a scale-selective MMI analysis of a histological section of the brain and a similar study of the birefringent fibrillar network of the histological section of the skeletal muscle revealed the formation of anharmonic distributions $C_{a,b}$ at all scales $a$ of the MHAT function window.

However, for the distributions of the MMI wavelet-coefficients $\{f_{44}(x, y); \Delta f(x, y)\}$, there are more "smooth" changes in the amplitudes of the wavelet-coefficients $W_{a,b}$ at different scales $a$ of the wavelet function $\Omega_{a,b}$ window. In accordance with this, the dynamics of differences between the values of statistical moments of the first to fourth orders that characterize such distributions changes—Table 3.6.

Thus, we have demonstrated the sensitivity at different scales of wavelet scanning $(Q_{a,b})$ of two-dimensional distributions of the values of the Mueller-matrix images $\{f_{44}(x, y); \Delta f(x, y)\}$ of polycrystalline networks of various types to changes in the linear and circular birefringence of such objects.

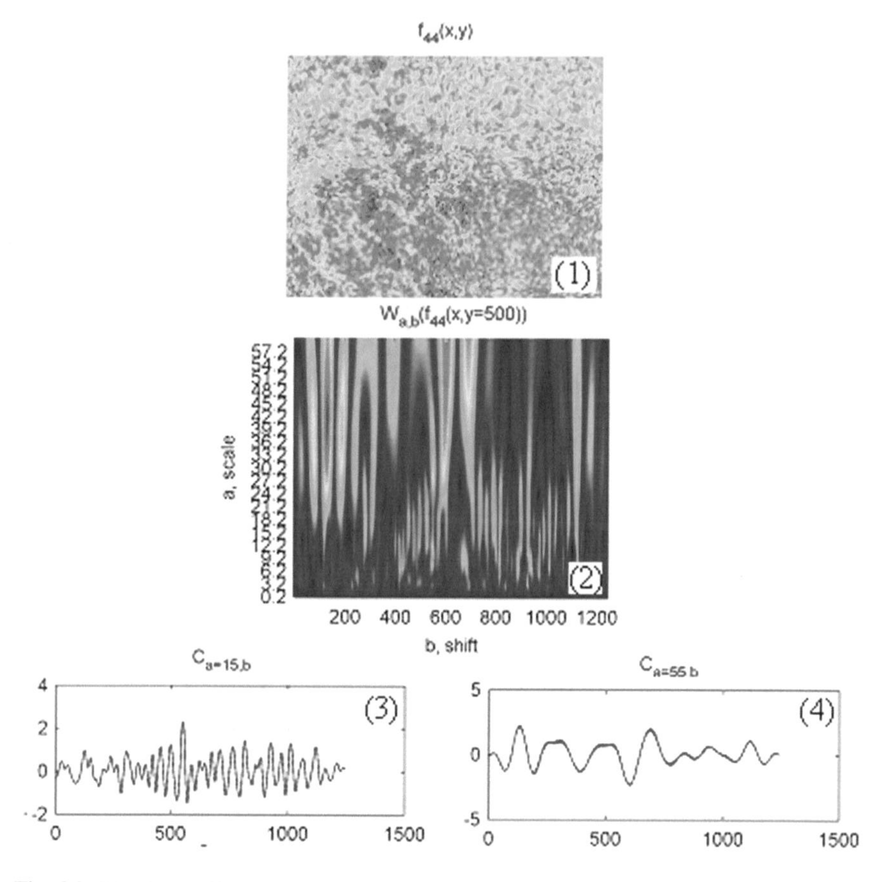


**Fig. 3.9** Wavelet-coefficients (2) and their different-scale ((3), (4)) cross sections of the MMI distribution $f_{44}$ (1) of the histological section of the brain

## 3.2.3 Diagnostic Application of Wavelet Analysis of Azimuthally Invariant Mueller-Matrix Images

The results of the wavelet analysis of azimuthally invariant Mueller-matrix images of polycrystalline networks of the myocardial histological sections of deceased due to mechanical asphyxiation and heart attack are shown in Figs. 3.11 and 3.12.

A comparative analysis of the results of the study of the MMI wavelet-coefficients $W_{a,b}(\Delta f)$ characterizing the manifestations of the optical activity of the polypeptide chains of myosin molecules revealed the greatest differences between them at the level of small scales $a_{\min}$ of the MHAT function. This fact is indicated by the large modulation depth of "small-scale" dependencies $C_{a=5,15,b}$. For the MMI $f_{44}$, which describes the linear birefringence of a network of spatially ordered myosin fibrils, such differences occur for "large-scale" dependencies $C_{a=80,b}$.

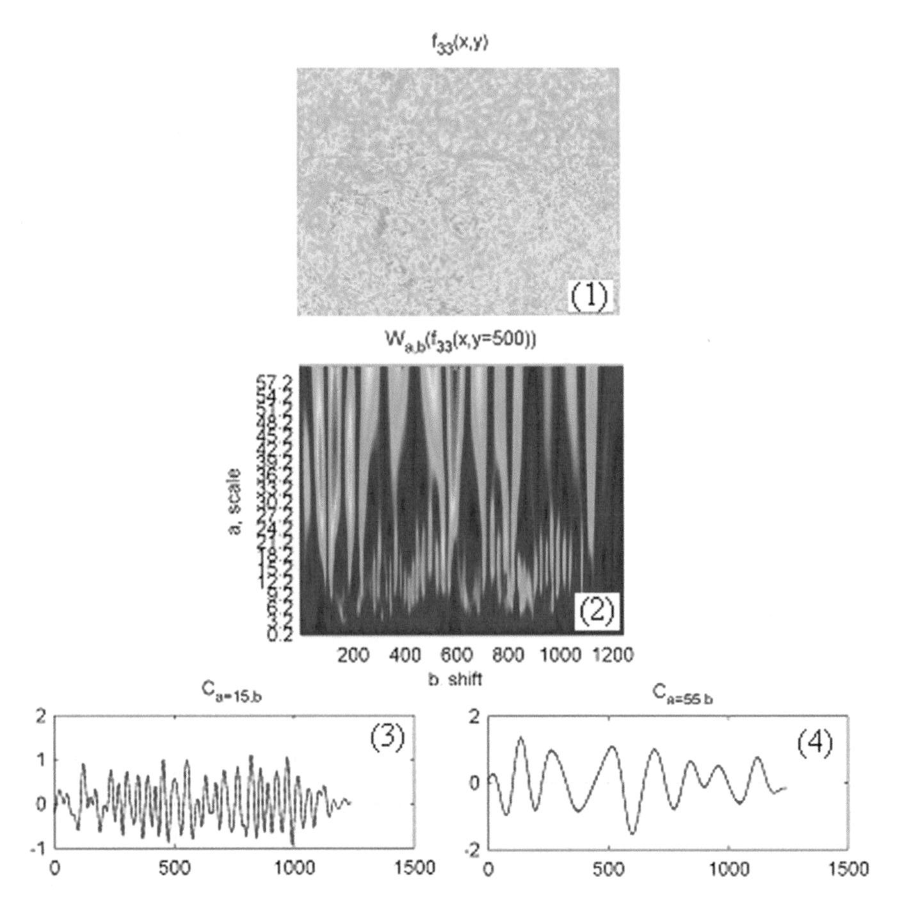


**Fig. 3.10** Wavelet-coefficients (2) and their different-scale ((3), (4)) cross sections of the MMI distribution $f_{33}$ (1) of the histological section of the brain

**Table 3.6** Statistical moments $Q_{i=1;2;3;4}$ that characterize the distribution of the values of the wavelet-coefficients of the MMI $f_{44}(m \times n)$ and ellipticity $\Delta f(m \times n)$ at the points of the microscopic image of the histological section of the brain

| Parameters | $f_{44}(m \times n)$ | $\Delta f(m \times n)$ | $f_{44}(m \times n)$ | $\Delta f(m \times n)$ | $f_{44}(m \times n)$ | $\Delta f(m \times n)$ |
|---|---|---|---|---|---|---|
| $a$ | 5 | | 25 | | 80 | |
| $Q_{i=1}$ | 0.03 | 0.015 | 0.025 | 0.021 | 0.018 | 0.038 |
| $Q_{i=2}$ | 0.22 | 0.21 | 0.16 | 0.15 | 0.12 | 0.14 |
| $Q_{i=3}$ | 0.24 | 0.11 | 0.29 | 0.08 | 0.34 | 0.05 |
| $Q_{i=4}$ | 0.15 | 0.59 | 0.37 | 0.47 | 0.59 | 0.32 |

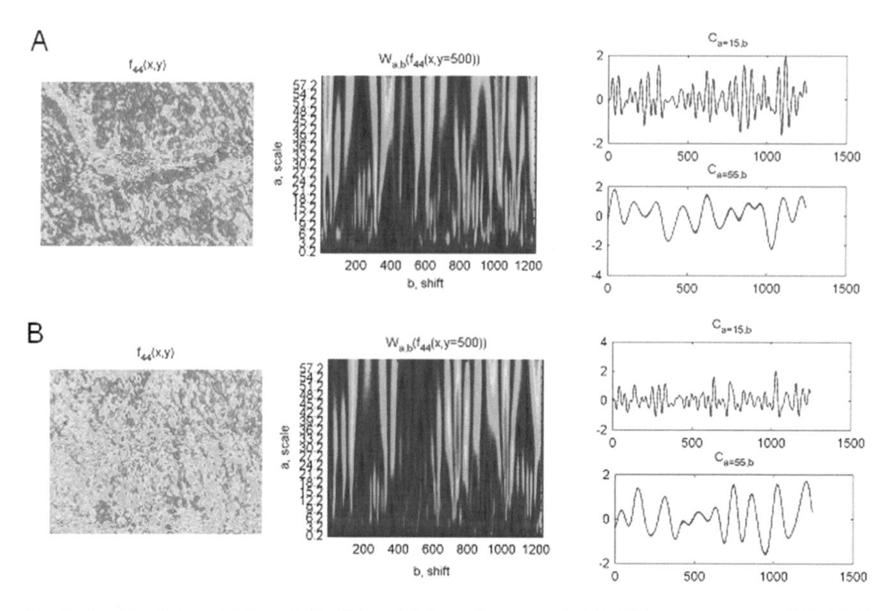


**Fig. 3.11** Wavelet-coefficients ((1), (3)) and their different-scale ((2), (4)) cross sections of the MMI distribution $f_{44}$ of the myocardium histological sections deceased due to mechanical asphyxiation ((1), (2)) and heart attack ((3) , (4))

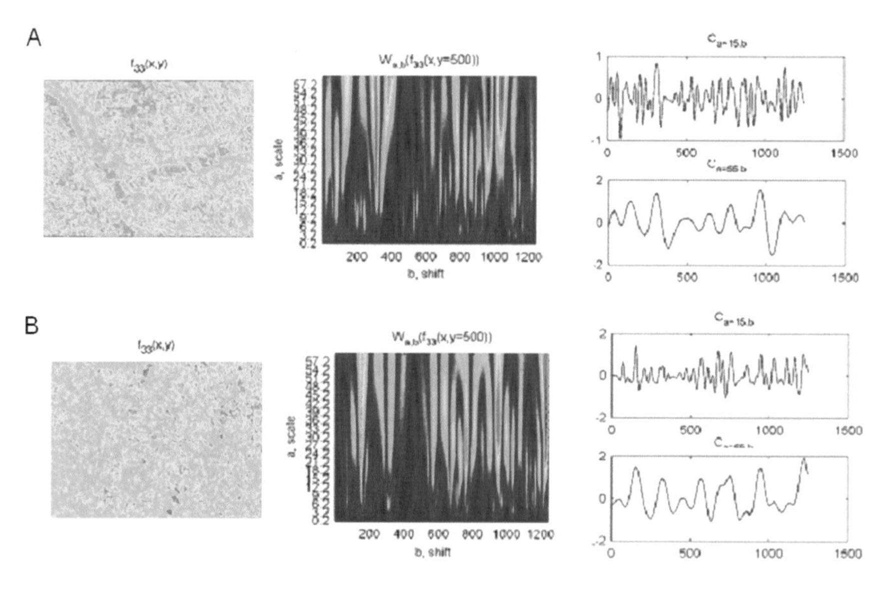


**Fig. 3.12** Wavelet-coefficients ((1), (3)) and their different-scale ((2), (4)) cross sections of the MMI distribution $\Delta f$ of the myocardium histological sections deceased due to mechanical asphyxiation ((1), (2)) and heart attack ((3) , (4))

**Table 3.7** Statistical moments $Q_{i=1;2;3;4}$ that characterize the distribution of the values of the wavelet-coefficients of the MMI of the myocardium histological sections deceased due to mechanical asphyxiation and heart attack (*), as well as the balanced accuracy of the method

| Parameters | $f_{44}(m \times n)$ | $f^*_{44}(m \times n)$ | $f_{44}(m \times n)$ | $f^*_{44}(m \times n)$ | $Ac(\%)$ | $Ac(\%)$ |
|---|---|---|---|---|---|---|
| $a$ | 5 | | 80 | | 5 | 80 |
| $Q_{i=1}$ | 0.04 ± 0.006 | 0.03 ± 0.005 | 0.018 ± 0.0024 | 0.014 ± 0.0019 | 55 | 58 |
| $Q_{i=2}$ | 0.31 ± 0.042 | 0.26 ± 0.033 | 0.12 ± 0.016 | 0.095 ± 0.001 | 58 | 62 |
| $Q_{i=3}$ | 0.27 ± 0.035 | 0.21 ± 0.029 | *0.34 ± 0.044* | *0.22 ± 0.031* | 68 | 78 |
| $Q_{i=4}$ | 0.17 ± 0.021 | 0.14 ± 0.018 | *0.59 ± 0.065* | *0.41 ± 0.053* | 70 | 80 |

The revealed patterns quantitatively characterizing the data are given in Tables 3.7 and 3.8.

The following differences (highlighted in italic) between the statistical parameters that characterize the distribution of values of different-scale cross sections of the values of wavelet maps $Q_{a,b}\{f_{44}(x, y); \Delta f(x, y)\}$ of polycrystalline networks of samples of myocardial histological sections $\begin{cases} a_{\min} \to \Delta M_3 = 1.39; \Delta M_4 = 1.25; \\ a_{\max} \to \Delta M_3 = 1.36; \Delta M_4 = 1.22. \end{cases}$.

At the same time, a high level of balanced accuracy ($Ac = 76$–$80\%$) of the proposed method is achieved, which is 5–10% higher than the diagnostic efficiency of the method of azimuthally invariant polarization mapping using wavelet analysis of azimuth distributions and polarization ellipticity at points of microscopic images of histological sections of the myocardium.

**Table 3.8** Statistical moments $Q_{i=1;2;3;4}$ that characterize the distribution of the values of the wavelet-coefficients of the MMI of the myocardium histological sections deceased due to mechanical asphyxiation ($\Delta f(m \times n)$) and heart attack ($\Delta f^*(m \times n)$), and the balanced accuracy of the method

| Parameters | $\Delta f(m \times n)$ | $\Delta f^*(m \times n)$ | $\Delta f(m \times n)$ | $\Delta f^*(m \times n)$ | $Ac(\%)$ | $Ac(\%)$ |
|---|---|---|---|---|---|---|
| $a$ | 5 | | 80 | | 5 | 80 |
| $Q_{i=1}$ | 0.025 ± 0.004 | 0.019 ± 0.003 | 0.085 ± 0.009 | 0.078 ± 0.0095 | 54 | 58 |
| $Q_{i=2}$ | 0.28 ± 0.034 | 0.23 ± 0.029 | 0.15 ± 0.022 | 0.12 ± 0.019 | 59 | 62 |
| $Q_{i=3}$ | *0.15 ± 0.022* | *0.24 ± 0.033* | 0.06 ± 0.008 | 0.09 ± 0.001 | 76 | 70 |
| $Q_{i=4}$ | *1.09 ± 0.12* | *0.87 ± 0.098* | 0.62 ± 0.077 | 0.54 ± 0.063 | 78 | 68 |

Note that this method of scale-selective analysis of the manifestations of the mechanisms of phase anisotropy opens up new possibilities not only in rapid, early diagnosis of inflammatory processes with latent (closed) course, but also in differentiating the severity of such conditions. However, all the results obtained by this method are not the result of direct (most valuable and reliable) measurements, but analytical wavelet processing of polarization and Mueller-matrix mapping data with subsequent large-scale selection of manifestations of optical anisotropy mechanisms. Therefore, an important next step is the development of methods for direct experimental detection and separation of manifestations of mechanisms of phase anisotropy of multiscale linearly and circularly birefringent polycrystalline networks of biological tissues and fluids.

## 3.3 A Brief Theory of the Method of Spatial-Frequency Filtration of Polarization-Inhomogeneous Microscopic Images of Histological Sections of Biological Tissues

It is known that any pathological conditions of human organs are accompanied by a transformation of both their biochemical composition and morphological structure. Such processes lead to a change in the relationship between the mechanisms of phase anisotropy of polycrystalline networks at different scales of their geometric dimensions. Therefore, for the further development of methods of azimuthally invariant polarization microscopy, the most urgent task is to develop techniques for experimental differentiation of the manifestations of linear and circular birefringence. One of the possible solutions may be spatial-frequency filtering of polarization-inhomogeneous object fields in the plane of microscopic images of histological sections of biological tissues. The first results in the application of this approach to the analysis of the structure of polycrystalline films of biological fluids in biomedical optics are given in [1, 2]. We are improving the developed Fourier–Stokes polarimetry method by applying it to a comprehensive comparative analysis of different frequency components of the azimuthally invariant polarization maps of microscopic and Mueller-matrix images of histological sections of tissues of human organs in different pathological conditions.

The main provisions of the optical model of the phase anisotropy of the polycrystalline structure of biological layers are considered and substantiated in detail [1, 2]. Therefore, we will present only brief, previously cited, theoretical information and some new relationships describing the azimuthally invariant method of spatial-frequency filtering of microscopic images of biological tissues.

During the propagation of a laser wave through a layer of biological tissue, due to the mechanisms of linear ($\gamma$—direction of the optical axis, $\phi$—phase shift) and circular ($\theta$—rotation of the plane of polarization) birefringence, amplitude-phase modulation occurs at the points of the microscopic image. In other words, the local parameters of the laser wave can be described by the partial Jones vector $\begin{pmatrix} E_x(\gamma, \phi, \theta) \\ E_y(\gamma, \phi, \theta) \end{pmatrix}$.

In [2], a relationship was found between the values of polarization $(\alpha, \beta)$ and amplitude-phase $(E_x(\gamma, \phi, \theta), E_y(\gamma, \phi, \theta)\ )$ parameters.

$$\alpha(\gamma, \varphi, \theta) = 0.5arctg\left(\frac{E_x(\gamma, \phi, \theta)E_y^*(\gamma, \phi, \theta) - E_y(\gamma, \phi, \theta)E_x^*(\gamma, \phi, \theta)}{E_x(\gamma, \phi, \theta)E_x^*(\gamma, \phi, \theta) - E_y(\gamma, \phi, \theta)E_y^*(\gamma, \phi, \theta)}\right); \tag{3.1}$$

$$\beta(\gamma, \phi, \theta) = 0.5 \arcsin\left(\frac{i\big(E_y(\gamma, \phi, \theta)E_x^*(\gamma, \phi, \theta) - E_x(\gamma, \phi, \theta)E_y^*(\gamma, \phi, \theta)\big)}{E_x(\gamma, \phi, \theta)E_x^*(\gamma, \phi, \theta) + E_y(\gamma, \phi, \theta)E_y^*(\gamma, \phi, \theta)}\right). \tag{3.2}$$

Here, $*$ denotes the complex conjugate.

The main idea of differentiating the frequency components in the azimuth and polarization elliptic distributions of this approach is to use spatial-frequency filtering. This method allows one to select either a low-frequency (with predominantly linear birefringence) or high-frequency (with predominantly circular birefringence) component in a polarization-inhomogeneous microscopic image of a histological section of biological tissue.

If one places a filtering $R(\Delta\nu, \Delta\mu)$ or vignette $R^{-1}(\Delta\nu, \Delta\mu)$ diaphragm in the central part of the Fourier plane, one can select a portion of the spatial-frequency structure of the Fourier spectrum that has developed under the predominant influence of linear $\hat{U}(\gamma, \phi, \nu, \mu)$ or circular $\dot{U}(\theta, \nu, \mu)$ birefringence of a polycrystalline film of biological fluid

$$\begin{cases} \hat{U}(\gamma, \phi, \nu, \mu) = R(\Delta\nu, \Delta\mu)U(\nu, \mu); \\ \dot{U}(\theta, \nu, \mu) = R^{-1}(\Delta\nu, \Delta\mu)U(\nu, \mu). \end{cases} \tag{3.3}$$

Here, $U_x$, $U_y$ are the Fourier distribution of complex amplitudes $E_x(\gamma, \phi, \theta)$ and $E_y(\gamma, \phi, \theta)$ with spatial frequencies $\nu = \frac{x}{\lambda f}$ and $\mu = \frac{y}{\lambda f}$.

The distribution of complex amplitudes $\begin{cases} \hat{E}_x(\gamma, \phi, x, y); \\ \dot{E}_x(\theta, x, y) \end{cases}$ and $\begin{cases} \hat{E}_y(\gamma, \phi, x, y); \\ \dot{E}_y(\theta, x, y) \end{cases}$ at the points of a microscopic image of a histological section of biological tissue can be restored using the inverse Fourier transform.

$$\begin{cases} \hat{E}_x(\gamma, \phi, x, y) = \frac{1}{i\lambda f} \int\limits_{-\infty}^{\infty} R(\Delta\nu, \Delta\mu)\hat{U}_x(\nu, \mu) \exp[i2\pi(x\Delta\nu + y\Delta\mu)]d\nu d\mu; \\ \hat{E}_y(\gamma, \phi, x, y) = \frac{1}{i\lambda f} \int\limits_{-\infty}^{\infty} R(\Delta\nu, \Delta\mu)\hat{U}_y(\nu, \mu) \exp[i2\pi(x\Delta\nu + y\Delta\mu)]d\nu d\mu; \\ \dot{E}_x(\theta, x, y) = \frac{1}{i\lambda f} \int\limits_{-\infty}^{\infty} R^{-1}(\Delta\nu, \Delta\mu)\dot{U}_x(\nu, \mu) \exp[i2\pi(x\Delta\nu + y\Delta\mu)]d\nu d\mu; \\ \dot{E}_y(\theta, x, y) = \frac{1}{i\lambda f} \int\limits_{-\infty}^{\infty} R^{-1}(\Delta\nu, \Delta\mu)\dot{U}_y(\nu, \mu) \exp[i2\pi(x\Delta\nu + y\Delta\mu)]d\nu d\mu. \end{cases}; \tag{3.4}$$

In our work, this technique was supplemented by azimuthally invariant polarization algorithms (see Chap. 2, paragraph 2.1.1, relation (2.1)– (2.3)) and Mueller-matrix mapping (see Chap. 2, paragraph 2.2.1, relation (2.4 ), (2.5)). Given the relationship between the values of the parameters of the Stokes vector and the amplitudes of the orthogonally polarized components of the laser wave

$$\begin{aligned} S_1 &= E_x E_x^* + E_y E_y^*; \\ S_2 &= E_x E_x^* - E_y E_y^*; \\ S_3 &= E_x E_y^* - E_y E_x^*; \\ S_4 &= i\left(E_y E_x^* - E_x E_y^*\right), \end{aligned} \tag{3.5}$$

the following relations can be written for the different frequency components of the distributions of polarization states at the points of the microscopic image

$$\hat{\alpha}^* = 0.5arctg\left(\frac{S_3}{S_2}\right) = 0.5arctg\left(\frac{\hat{E}_x\hat{E}_y^* - \hat{E}_y\hat{E}_x^*}{\hat{E}_x\hat{E}_x^* - \hat{E}_y\hat{E}_y^*}\right); \tag{3.6}$$

$$\dot{\alpha}^* = 0.5arctg\left(\frac{S_3}{S_2}\right) = 0.5arctg\left(\frac{\dot{E}_x\dot{E}_y^* - \dot{E}_y\dot{E}_x^*}{\dot{E}_x\dot{E}_x^* - \dot{E}_y\dot{E}_y^*}\right); \tag{3.7}$$

$$\hat{\beta}^* = 0,5\arcsin(S_4) = 0.5\arcsin i\left(\hat{E}_y\hat{E}_x^* - \hat{E}_x\hat{E}_y^*\right); \tag{3.8}$$

$$\dot{\beta}^* = 0.5\arcsin(S_4) = 0.5\arcsin i\left(\dot{E}_y\dot{E}_x^* - \dot{E}_x\dot{E}_y^*\right). \tag{3.9}$$

## 3.4 Fourier Analysis of Azimuthally Invariant Distributions of Polarization Parameters of Microscopic Images of Biological Tissues

In the framework of applications of the spatial-frequency filtering mode in the system of azimuthally invariant polarization mapping of microscopic images of histological sections of biological tissues, we investigated the possibility of differentiating the changes in birefringence of the myocardium deceased due to mechanical asphyxiation (Figs. 3.13 and 3.14) and heart attack (Figs. 3.15 and 3.16).

A comparative analysis of the obtained data on the distribution of polarization parameters due to changes in birefringence at different scales of optically anisotropic protein networks revealed the following differences. For the polarization azimuth map, there is a higher level of circular birefringence of myosin polypeptide chains of the deceased due to mechanical asphyxiation (see Figs. 3.13 and 3.15). For the

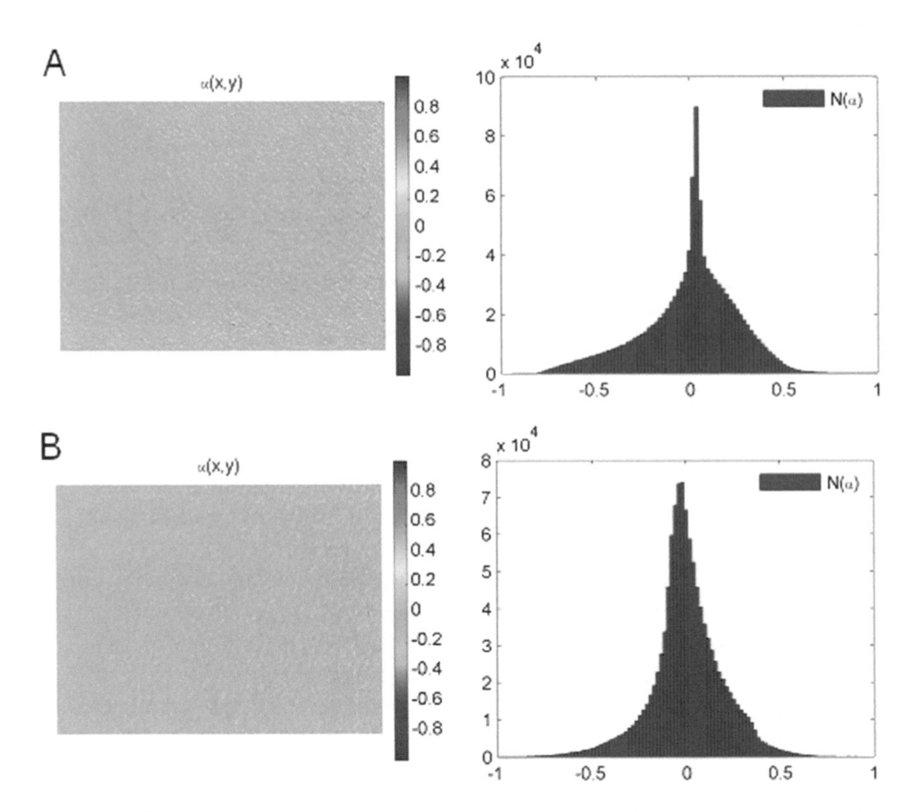


**Fig. 3.13** "High-frequency" two-dimensional components and histograms of the distribution of polarization azimuth values in microscopic images of the myocardium histological sections deceased due to mechanical asphyxiation (A) and heart attack (B)

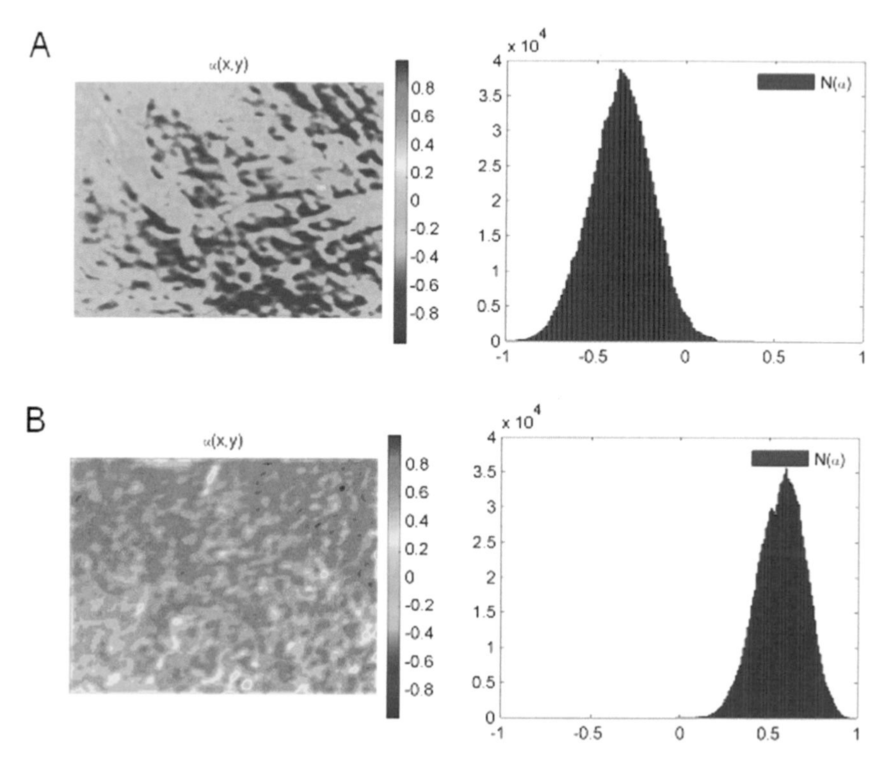


**Fig. 3.14** "Low-frequency" two-dimensional components and histograms of the distribution of polarization azimuth values in microscopic images of the myocardium histological sections deceased due to mechanical asphyxiation (A) and heart attack (B)

polarization ellipticity map, a decrease in the linear birefringence of myosin networks of the myocardium that deceased of a heart attack (see Figs. 3.14 and 3.16).

Quantitatively, the polarization distributions (relation (3.6)–(3.9)) are illustrated by a set of statistical moments $Q_{i=1;2;3;4}$, the values of which are given in Tables 3.9 and 3.10.

A comparative analysis of the data revealed the following intergroup differences in the values of the statistical moments $Q_{i=1;2;3;4}$ characterizing the phase distributions $\alpha^*(x, y) \Rightarrow \{\Delta Q_3 = 1.37 \Delta Q_4 = 1.44\}$; $\beta^*(x, y) \Rightarrow \{\Delta Q_3 = 1.35; \Delta Q_4 = 1.29\}$. Analysis of the diagnostic effectiveness of the method of azimuthally invariant polarization mapping with spatial-frequency filtering of microscopic images of histological sections of the myocardium revealed a high level of balanced accuracy $Ac(\alpha) = 80\% - 82\%$, $Ac(\beta) = 78$--$82\%$. This is 5–15% higher than the azimuthally invariant polarization-correlation microscopy using scale-selective wavelet analysis.

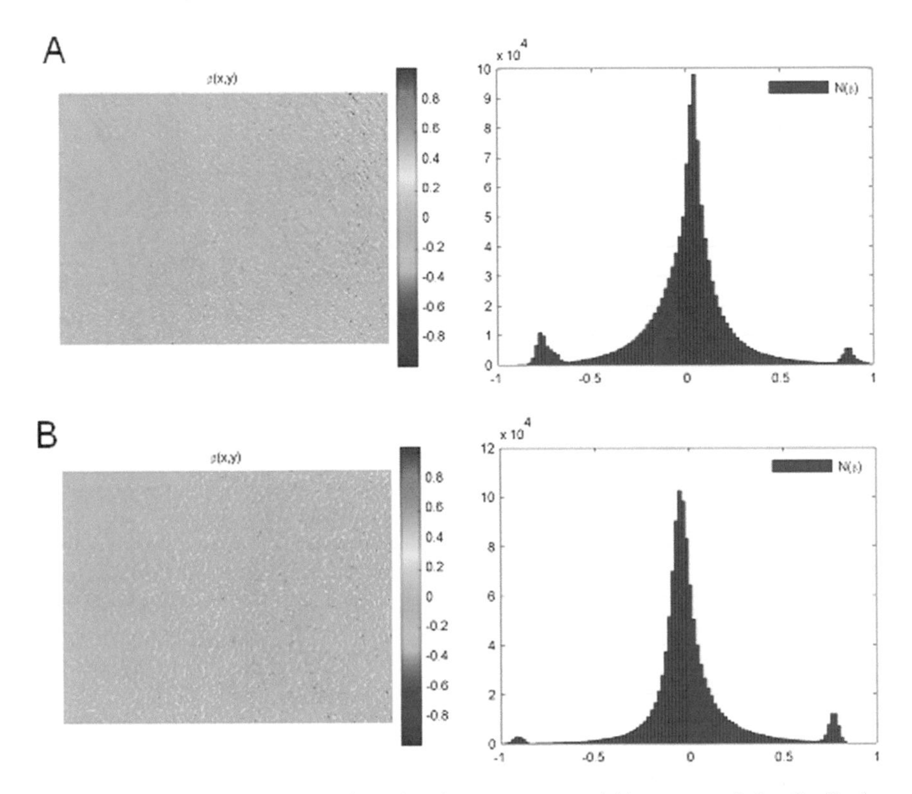


**Fig. 3.15** "High-frequency" two-dimensional components and histograms of the distribution of polarization ellipticity values in microscopic images of the myocardial histological sections deceased due to mechanical asphyxiation (A) and heart attack (B)

## 3.5 Fourier Analysis of Azimuthally Invariant Mueller-Matrix Images of Biological Tissues

The next step in the study was the azimuthally invariant Mueller-matrix mapping of optically anisotropic networks of histological sections of the myocardium deceased due to mechanical asphyxiation (Figs. 3.17 and 3.18) and heart attack (Figs. 3.19 and 3.20).

A comparative analysis of the obtained distributions of MMI, which are characterized by changes in birefringence at different scales of optically anisotropic protein networks, revealed the following differences. For the Mueller-matrix image $\Delta f$—a higher level of circular birefringence of myosin polypeptide chains of the myocardium that deceased due to mechanical asphyxiation. For the Mueller-matrix image $f_{44}$—a decrease in the linear birefringence of myosin networks of the myocardium that deceased of a heart attack.

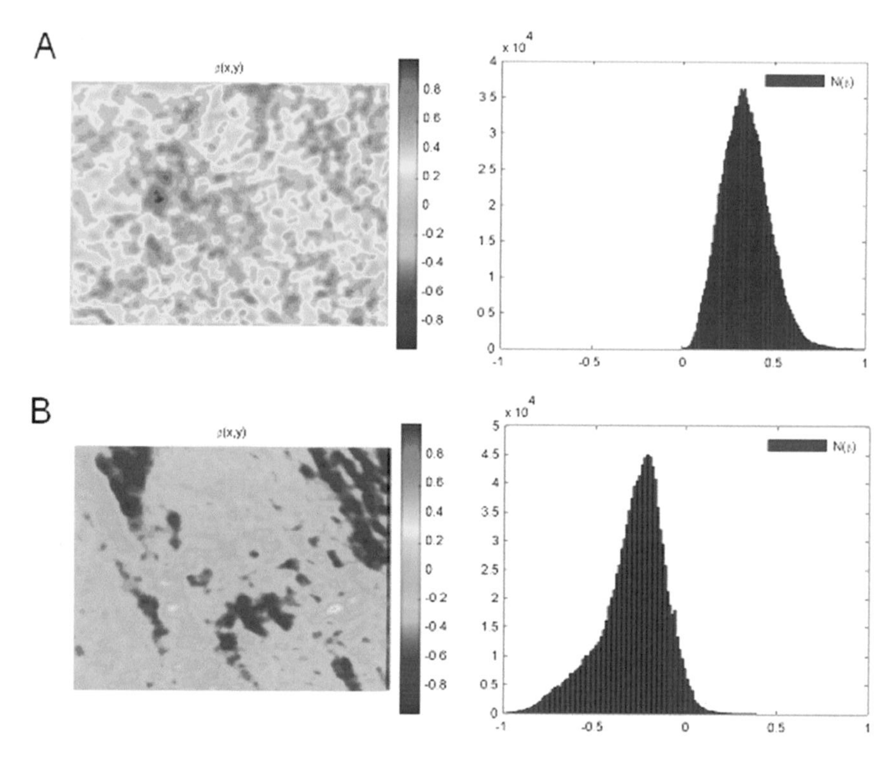


**Fig. 3.16** " Low-frequency" two-dimensional components and histograms of the distribution of polarization ellipticity values in microscopic images of the myocardium histological sections deceased due to mechanical asphyxiation (A) and heart attack (B)

**Table 3.9** Statistical moments $Q_{i=1;2;3;4}$ that characterize the distribution of polarization azimuth values of microscopic images of histological sections of the myocardium of both groups

| Parameters | $\alpha(m \times n)$ | $\alpha^*(m \times n)$ | $\alpha(m \times n)$ | $\alpha^*(m \times n)$ | $Ac(\%)$ | $Ac(\%)$ |
|---|---|---|---|---|---|---|
| $\nu$ | High-frequency component ($\nu^*$) | | "Low-frequency" component ($\nu^{**}$) | | $\nu^*$ | $\nu^{**}$ |
| $Q_{i=1}$ | 0.16 ± 0.024 | 0.12 ± 0.017 | 0.56 ± 0.072 | 0.51 ± 0.063 | 61 | 54 |
| $Q_{i=2}$ | 0.08 ± 0.001 | 0.07 ± 0.009 | 0.15 ± 0.021 | 0.12 ± 0.018 | 54 | 58 |
| $Q_{i=3}$ | *0.38 ± 0.044* | *0.52 ± 0.068* | 1.47 ± 0.19 | 1.79 ± 0.24 | 80 | 72 |
| $Q_{i=4}$ | *1.24 ± 0.19* | *1.79 ± 0.21* | 1.18 ± 0.15 | 1.36 ± 0.19 | 82 | 67 |

**Table 3.10** Statistical moments $Q_{i=1;2;3;4}$ that characterize the distribution of polarization ellipticity values of microscopic images of histological sections of the myocardium

| Parameters | $\beta(m \times n)$ | $\beta^*(m \times n)$ | $\beta(m \times n)$ | $\beta^*(m \times n)$ | $Ac(\%)$ | $Ac(\%)$ |
|---|---|---|---|---|---|---|
| $\nu$ | High-frequency component ($\nu^*$) | | "Low-frequency" component ($\nu^{**}$) | | $\nu^*$ | $\nu^{**}$ |
| $Q_{i=1}$ | 0.078 ± 0.009 | 0.063 ± 0.008 | 0.31 ± 0.048 | 0.28 ± 0.036 | 54 | 56 |
| $Q_{i=2}$ | 0.13 ± 0.017 | 0.11 ± 0.015 | 0.14 ± 0.021 | 0.11 ± 0.014 | 56 | 58 |
| $Q_{i=3}$ | 0.65 ± 0.088 | 0.71 ± 0.089 | *1.32 ± 0.18* | *1.78 ± 0.24* | 67 | 78 |
| $Q_{i=4}$ | 1.86 ± 0.25 | 1.98 ± 0.29 | *0.63 ± 0.089* | *0.49 ± 0.062* | 69 | 82 |

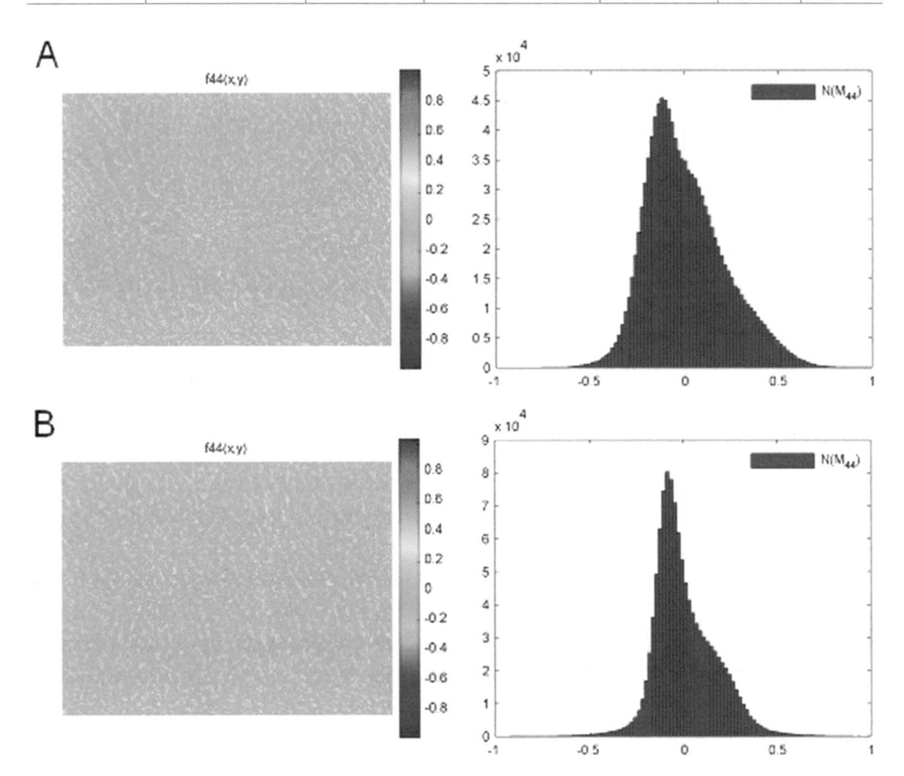


**Fig. 3.17** "High-frequency" two-dimensional components and histograms of the distribution of the MMI values $f_{44}$ of the myocardium histological sections deceased due to mechanical asphyxiation (A) and heart attack (B)

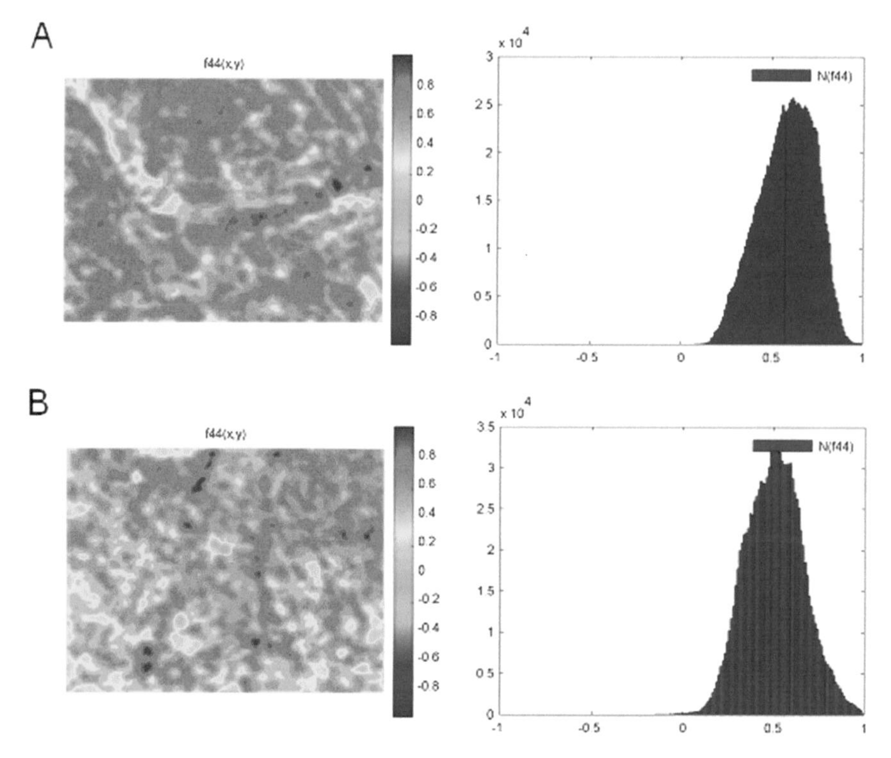


**Fig. 3.18** "Low-frequency" two-dimensional components and histograms of the distribution of the MMI values $f_{44}$ of the myocardium histological sections deceased due to mechanical asphyxiation (A) and heart attack (B)

A quantitative distribution of the values of the elements of the Mueller-matrix invariants is illustrated by a set of statistical moments $Q_{i=1;2;3;4}$, the values of which are given in Tables 3.11 and 3.12.

A comparative analysis of the obtained data revealed an increase (by 35–45%) compared with the data of polarization mapping (see Tables 3.9 and 3.10) of inter-group differences in the values of statistical moments $Q_{i=1;2;3;4}$ characterizing the distribution of azimuthally invariant elements of the Mueller matrix and their combinations.

$\Delta f(x, y) \Rightarrow \{\Delta Q_3 = 1.54 \Delta Q_4 = 1.47\}$;

$$f_{44}(x, y) \Rightarrow \{\Delta Q_3 = 1.5; \Delta Q_4 = 1.47\}.$$

A comparative analysis of the diagnostic efficiency of the methods of azimuthally invariant Mueller matrix and polarization mapping with spatial-frequency filtering of microscopic images of histological sections of the myocardium revealed an increase in the level of balanced accuracy $Ac(\Delta f) = 84\% - 86\%$, $Ac(f_{44}) = 82\% - 86\%$.

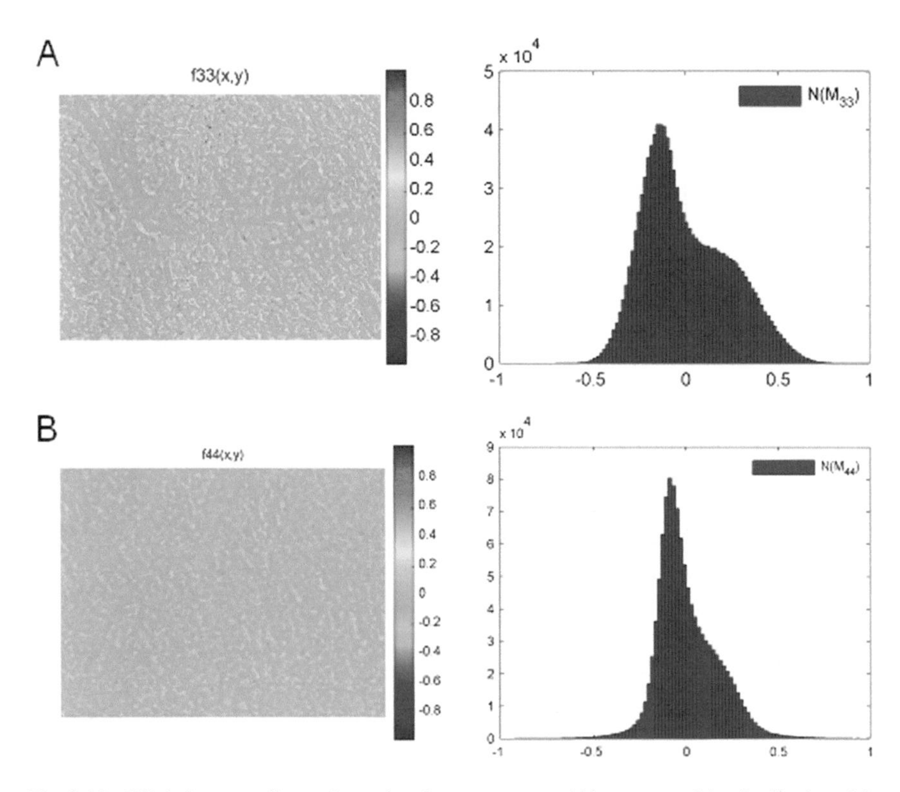


**Fig. 3.19** "High-frequency" two-dimensional components and histograms of the distribution of the MMI values $f_{33}$ of the myocardium histological sections deceased due to mechanical asphyxiation (A) and heart attack (B)

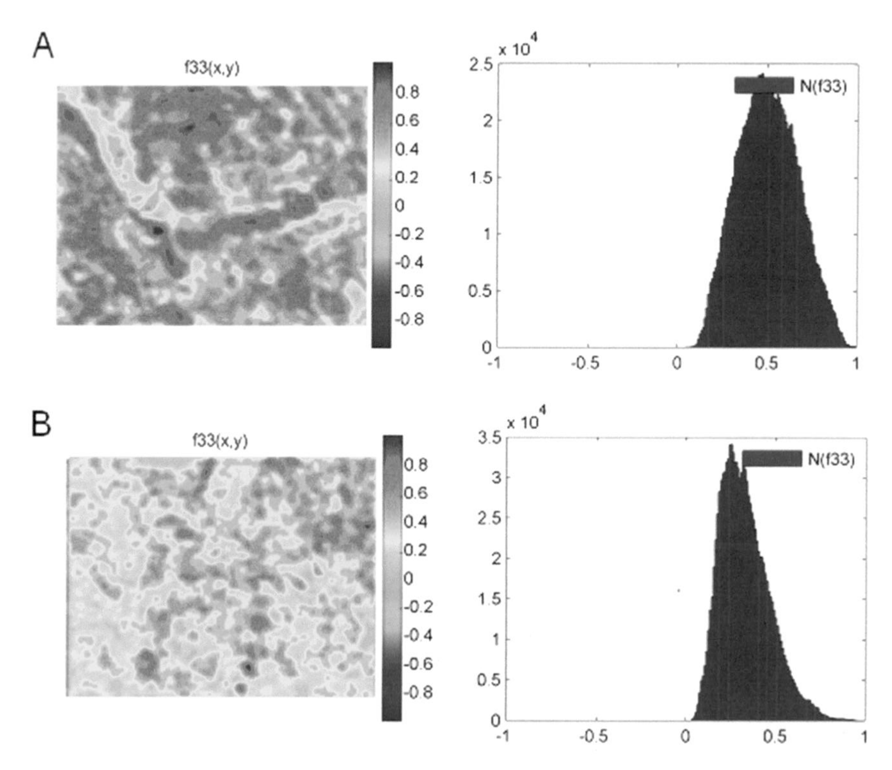


**Fig. 3.20** "Low-frequency" two-dimensional components and histograms of the distribution of the MMI values $f_{33}$ of the myocardium histological sections deceased due to mechanical asphyxiation (A) and heart attack (B)

**Table 3.11** Statistical moments $Q_{i=1;2;3;4}$ that characterize the distribution of MMI values at the points of the "high-frequency" and "low-frequency" component of the Mueller-matrix images of the myocardium histological sections deceased due to mechanical asphyxiation $f_{44}(m \times n)$ and heart attack $f_{44}^{*}(m \times n)$

| Parameters | $f_{44}(m \times n)$ | $f_{44}^{*}(m \times n)$ | $f_{44}(m \times n)$ | $f_{44}^{*}(m \times n)$ | $Ac(\%)$ | $Ac(\%)$ |
|---|---|---|---|---|---|---|
| $\nu$ | High-frequency component ($\nu^{*}$) | | "Low-frequency" component ($\nu^{**}$) | | $\nu^{*}$ | $\nu^{**}$ |
| $Q_{i=1}$ | 0.11 ± 0.015 | 0.15 ± 0.019 | 0.36 ± 0.048 | 0.44 ± 0.061 | 65 | 67 |
| $Q_{i=2}$ | 0.12 ± 0.017 | 0.14 ± 0.018 | 0.18 ± 0.024 | 0.22 ± 0.031 | 60 | 62 |
| $Q_{i=3}$ | 0.45 ± 0.058 | 0.31 ± 0.045 | *0.33 ± 0.049* | *0.22 ± 0.033* | 75 | 82 |
| $Q_{i=4}$ | 0.81 ± 0.099 | 0.62 ± 0.083 | *0.28 ± 0.037* | *0.19 ± 0.024* | 77 | 86 |

**Table 3.12** Statistical moments $Q_{i=1;2;3;4}$ that characterize the distribution of MMI values at the points of the "high-frequency" and "low-frequency" components of the Mueller-matrix images of the myocardium histological sections deceased due to mechanical asphyxiation $\Delta f(m \times n)$ and heart attack $\Delta f^*(m \times n)$

| Parameters | $\Delta f(m \times n)$ | $\Delta f^*(m \times n)$ | $\Delta f(m \times n)$ | $\Delta f^*(m \times n)$ | $Ac(\%)$ | $Ac(\%)$ |
|---|---|---|---|---|---|---|
| $\nu$ | High-frequency component ($\nu^*$) | | "Low-frequency" component ($\nu^{**}$) | | $\nu^*$ | $\nu^{**}$ |
| $Q_{i=1}$ | 0.16 ± 0.022 | 0.13 ± 0.019 | 0.09 ± 0.01 | 0.07 ± 0.009 | 62 | 58 |
| $Q_{i=2}$ | 0.13 ± 0.016 | 0.11 ± 0.014 | 0.12 ± 0.017 | 0.106 ± 0.015 | 61 | 63 |
| $Q_{i=3}$ | *0.57 ± 0.071* | *0.88 ± 0.11* | 0.69 ± 0.088 | 0.57 ± 0.073 | 86 | 74 |
| $Q_{i=4}$ | *1.21 ± 0.18* | *1.78 ± 0.23* | 0.87 ± 0.11 | 0.72 ± 0.089 | 84 | 72 |

## References

1. A. Doronin, C. Macdonald, I. Meglinski, Propagation of coherent polarized light in turbid highly scattering medium. J. Biomed. Opt. **19**(2), 025005 (2014)
2. A. Doronin, A. Radosevich, V. Backman, I. Meglinski, Two electric field Monte Carlo models of coherent backscattering of polarized light. J. Opt. Soc. America A **31**(11), 2394 (2014)

# Chapter 4
# Polarization Correlometry of Microscopic Images of Polycrystalline Networks Biological Layers

## 4.1 Polarization Correlometry of Microscopic Images of Biological Layers

### 4.1.1 *Brief Theory of the Method*

This section presents the results of the practical application of the theoretical correlation approach for combining the polarization [1–14] and correlation [15–23] approaches to the analysis of polarization-inhomogeneous fields by introducing a polarization-correlation parameter—a complex degree of mutual polarization [21].

In works [22, 23], an expression of the CDMP value was found for two points $(r_1, r_2)$ of a microscopic image of a histological section of biological tissue

$$V(r_1, r_2) = \frac{\left(U_x(r_1)U_x(r_2) - U_y(r_1)U_y(r_2)\right)^2 + 4U_x(r_1)U_x(r_2)U_y(r_1)U_y(r_2)\exp\left(i\left(\delta_y(r_2) - \delta_x(r_1)\right)\right)}{\left(U_x^2(r_1) + U_y^2(r_1)\right)\left(U_x^2(r_2) + U_y^2(r_2)\right)}. \quad (4.1)$$

Here, $U_x$ and $U_y$ are the orthogonal components of the amplitude of the laser wave in the plane of the microscopic image, $\delta_x$ and $\delta_y$ are the corresponding phase values of such components.

The main characteristic values of the parameter $V$ for points $(r_1, r_2)$ with different types of polarization are given in Table 4.1.

From the data in Table 4.1, it follows that the magnitude of the complex degree of mutual polarization $V(r_1, r_2)$ can be used as a diagnostic parameter for evaluating the coordinate polarization-inhomogeneous structure of microscopic images of biological objects.

For the purpose of the experimental application of this parameter, we determine the real ($\mathrm{Re}\{V\}$) part of the complex degree of mutual polarization $V$:

 
I. Meglinski et al., *Shedding the Polarized Light on Biological Tissues*,
SpringerBriefs in Applied Sciences and Technology,
https://doi.org/10.1007/978-981-10-4047-4_4

**Table 4.1** Value of the complex degree of mutual polarization $V(r_1, r_2)$ of two points $(r_1, r_2)$ with different types of light vibrations

| $U(r_1)$ | $U(r_2)$ | $V(r_1, r_2)$ |
|---|---|---|
| $U_x(r_1)$; $U_x(r_1) + U_y(r_1)\exp(i\,\delta)$ | $U_x(r_2)$; $U_x(r_2) + U_y(r_2)\exp(i\,\delta)$ | 1 |
| $U_x(r_1)$ | $U_x(r_2) + U_y(r_2)$ | $\frac{U_x^2(r_2)}{U_x^2(r_2)+U_y^2(r_2)}$ |
| $U_x(r_1)$ | $U_x(r_2) + iU_y(r_2)$ | 0.5 |
| $U_x(r_1)$; $U_x(r_1) + U_y(r_1)\exp(i\,\delta)$ | $U_y(r_2)$; $U_x(r_2) - U_y(r_2)\exp(i\,\delta)$ | 0 |

$$\mathrm{Re}\{V(r_1, r_2)\} = \frac{\left(U_x(r_1)U_x(r_2) - U_y(r_1)U_y(r_2)\right)^2 + 4U_x(r_1)U_x(r_2)U_y(r_1)U_y(r_2)\cos\delta_{12}}{\left(U_x^2(r_1) + U_y^2(r_1)\right)\left(U_x^2(r_2) + U_y^2(r_2)\right)}; \quad (4.2)$$

Chapter 2, paragraph 2.3.1 [relation (2.6) (2.7)], provides a methodology for experimental measurement of the coordinate distribution of the values of the CDMP modulus at the points of a microscopic image of a histological section of a biological tissue.

### *4.1.2 CDMP Mapping of Microscopic Images of Biological Layers with Ordered Architectonics*

Figure 4.1 shows the results of an experimental study of the coordinate distributions of the values of the CDMP modulus (fragment 1) of a histological section of skeletal muscle tissue, as well as a histogram of the distribution of random values of the CDMP modulus (fragment 2).

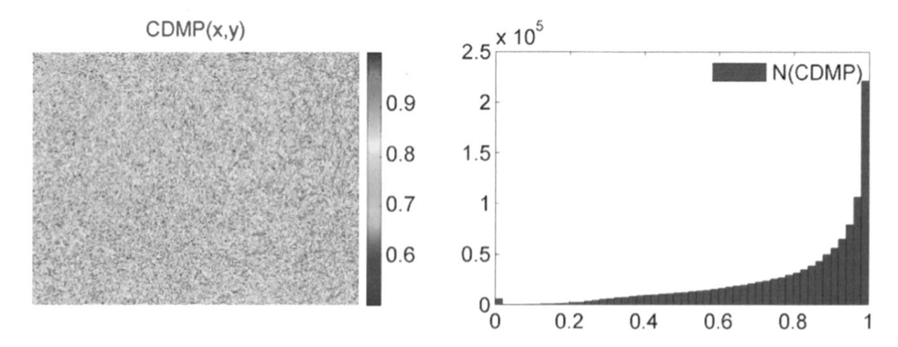


**Fig. 4.1** CDMP map (left side) and a histogram (right side) of the distribution of CDMP modulus values in a microscopic image of a histological section of skeletal muscle

**Table 4.2** Statistical moments $Q_{i=1;2;3;4}$ that characterize the coordinate distribution of the values of the CDMP modulus of the microscopic image of a histological section of the skeletal muscle

| $Q_{i=1;2;3;4}$ | CDMP |
|---|---|
| $Q_{i=1}$ | 0.89 |
| $Q_{i=2}$ | 0.14 |
| $Q_{i=3}$ | 0.71 |
| $Q_{i=4}$ | 1.36 |

From the obtained data, it is seen that the optical anisotropy of the histological section of the skeletal muscle with a spatially structured fibrillar network are manifested in the formation of a polarization-inhomogeneous microscopic image. From a physical point of view, the main factor in this is the presence of a certain spectrum of orientations of the optical axes and phase shifts introduced by birefringent fibrils with different geometric cross sections. On the other hand, the polycrystalline myosin network of the skeletal muscle is rather spatially ordered. Based on this, the structure of the histogram of the distribution of random values of the CDMP modulus becomes clear—asymmetric dependences with a fairly small dispersion and a clearly defined extremum.

Quantitatively, the results of a statistical analysis of the coordinate distributions of CDMP values of optically thin layers of the myocardium and cerebrospinal fluid film are shown in Table 4.2.

Analysis of the full set of statistical moments that characterize the coordinate distributions of the values of the CDMP modulus of a microscopic image of a histological section of a biological tissue with a spatially ordered polycrystalline network revealed the most sensitive of them. These include higher-order statistical moments $Q_{i=3;4}(V)$ characterizing the skewness and sharpness of the peak of the histograms of the distributions of the values of the CDMP modulus of microscopic images of such objects.

### *4.1.3 CDMP Mapping of Microscopic Images of Biological Layers with Disordered Architectonics*

Figure 4.2 illustrates the results of an experimental study of coordinate distributions of values of the CDMP modulus (left side) of a histological section of brain tissue, as well as a histogram of the distribution of random values of the CDMP modulus (right side).

A comparative analysis of the coordinate distributions of the values of the CDMP modulus of microscopic images of histological sections of the myocardium (see Fig. 4.1) and brain tissue (see Fig. 4.2) revealed significant differences. In particular, the structure of the polarization-inhomogeneous microscopic image is more complex in the sense of increasing the interval of random changes in the values of the CDMP modulus. The latter, in turn (see Table 4.1), are determined by the degree of similarity,

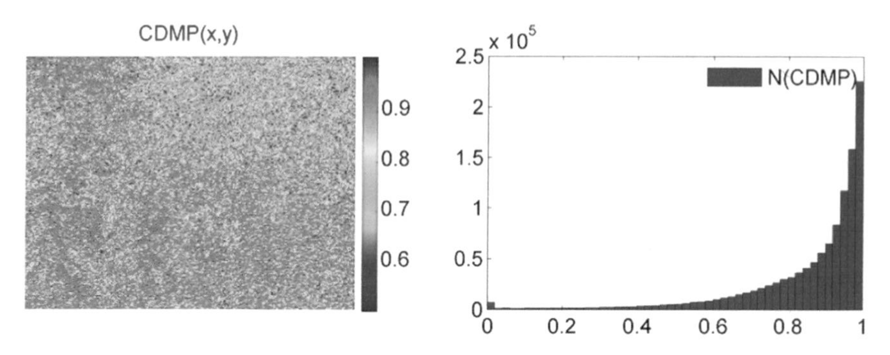


**Fig. 4.2** CDMP map (left side) and a histogram (right side) of the distribution of values of the CDMP modulus in a microscopic image of a histological section of the brain

**Table 4.3** Statistical moments $Q_{i=1;2;3;4}$ that characterize the coordinate distribution of values of the CDMP modulus of a microscopic image of a histological section of the brain

| $Q_{i=1;2;3;4}$ | CDMP |
|---|---|
| $Q_{i=1}$ | 0.74 |
| $Q_{i=2}$ | 0.21 |
| $Q_{i=3}$ | 0.57 |
| $Q_{i=4}$ | 0.83 |

or, conversely, by differences in polarization states at the points of a microscopic image of a histological section of brain tissue.

The discovered feature can be associated with a wide range of orientations of the optical axes of local fibrils of a given biological tissue. This morphological structure of the optically anisotropic fibrillar network makes it possible to form more "diverse" polarization states in different coordinates of the microscopic image. Based on this, a greater level of dispersion of the histogram of the distribution of random values of the CDMP modulus (see Fig. 4.2, right part) becomes clear, while reducing its skewness and peak sharpness - Table 4.3.

A comparative analysis of the magnitude of the statistical moments of the first to fourth orders (see Tables 4.2 and 4.3) characterizing the coordinate distribution of the values of the CDMP modulus of polycrystalline networks of histological sections of biological tissues of both types was found:

- Various types of polycrystalline networks of real biological tissues are characterized by an individual set of statistical moments $Q_{i=1;2;3;4}(V)$ characterizing the distribution of the degree of correlation consistency of azimuth and ellipticity of polarization at different points of the microscopic image of the object.
- The most sensitive to changes in the polarization inhomogeneity of laser fields in the plane of microscopic images were all statistical moments $Q_{i=1;2;3;4}(V)$.

- The following trends in the magnitude of statistical moments of the first to fourth orders that characterize the polarization-correlation structure of microscopic images with a birefringent fibrillar network disordered in the directions of the optical axes—$Q_1 \downarrow$; $Q_2 \uparrow$; $Q_3 \downarrow$; $Q_4 \downarrow$.

The results and certain patterns of formation of CDMP distributions of networks of spatially structured fibrillar networks of biological crystals were the basis for the development of a method for differentiating changes in optical anisotropy due to necrotic changes in the myocardium.

### 4.1.4 CDMP Mapping Diagnostic Features

This paragraph contains the results of studies on the possibility of polarization-correlation differentiation of microscopic images of histological sections of the myocardium that deceased due to mechanical asphyxiation and heart attack. Figure 4.3 shows a series of coordinate distributions of the values of the modulus of the polarization-correlation parameter CDMP (left parts) and histograms of their

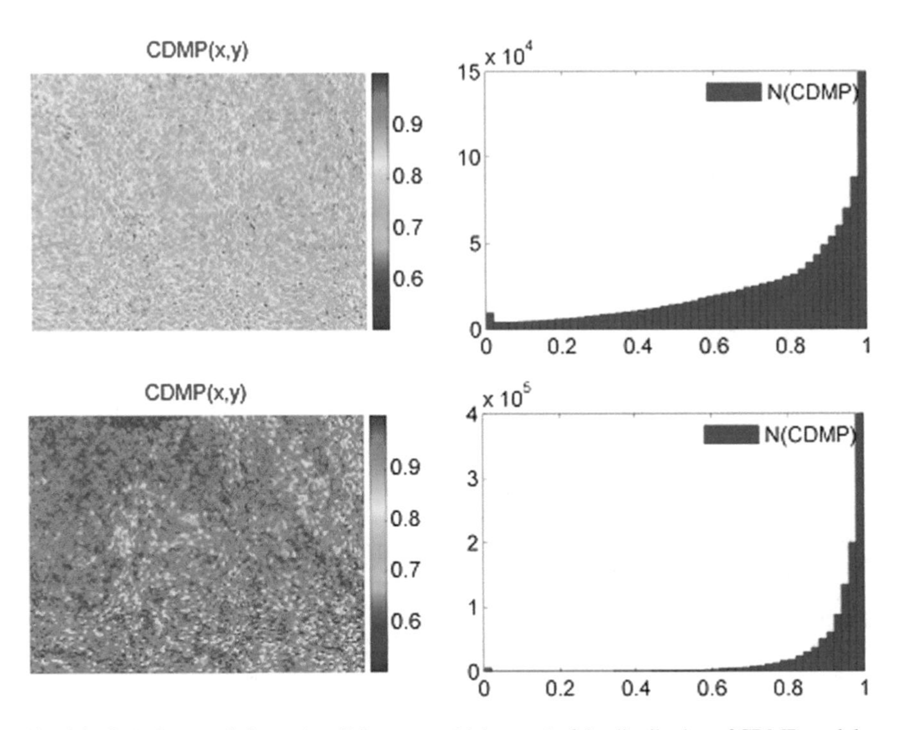


**Fig. 4.3** CDMP maps (left parts) and histograms (right parts) of the distribution of CDMP modulus values in microscopic images of histological sections of the myocardium that deceased from mechanical asphyxiation (top line) and heart attack (bottom line)

random values (right parts). A comparison of the statistical parameters $Q_{i=1;2;3;4}(V)$ characterizing the distribution of the values of the CDMP modulus of microscopic images of histological sections of the myocardium of both types revealed the following differences for mechanical asphyxiation compared with the case of a heart attack:

- Decrease in average ($Q_1(V)\ \downarrow\uparrow$), skewness ($Q_3(V)\ \downarrow\uparrow$) and kurtosis ($Q_4(V)\downarrow\uparrow$);
- Increase of dispersion ($Q_2(V)\ \uparrow$), which characterizes the histograms of the distribution of the value of the CDMP modulus, for the image of the sample of the myocardium deceased due to mechanical asphyxiation.

Physically obtained results can be associated with the "destruction" of birefringence of the fibrillar network of the myocardium, which deceased due to a heart attack. As a result, the probability of formation of CDMP modulus values other than extreme decreases $V = 1$. As a result, the dispersion decreases and, on the contrary, the average, skewness, and kurtosis increase, characterizing the distribution of this correlation parameter of the microscopic image of the histological section of the necrotic altered myocardium.

The extension of the range of variation of random values of the CDMP modulus in the plane of the microscopic image of the myocardium histological section deceased due to mechanical asphyxiation is associated with a high level of birefringence of the myosin network of this sample—$Q_1\downarrow$; $Q_2\uparrow$; $Q_3\downarrow$; $Q_4\downarrow$.

The results of the comparative analysis of the averaged (within the respective groups of samples) values of the statistical moments $Q_{i=1;2;3;4}$ characterizing the two-dimensional distribution of the values of the CDMP modulus are given in Table 4.4.

The data obtained revealed the most sensitive ($\Delta Q = \max$) to changes in the polarization structure of microscopic images of polycrystalline myocardial networks—statistical moments $Q_1$, $Q_2$.

The following quantitative criteria for intergroup differences in the values of statistical moments of the first–second orders characterizing the distribution of the modulus of the polarization-correlation parameter CDMP of microscopic images of histological sections of the myocardium of both groups have been established—$V(m\times n) \Leftrightarrow \Delta Q_1 = 1,35$; $\Delta Q_2 = 1,69$.

**Table 4.4** Statistical moments $Q_{i=1;2;3;4}$ that characterize the distribution of the values of the CDMP modulus of the myocardium histological sections deceased due to mechanical asphyxiation ($V$) and heart attack ($V^*$)

| $Q_{i=1;2;3;4}$ | $V$ | $V^*$ | $Ac(\%)$ |
|---|---|---|---|
| $Q_{i=1}$ | $0.68 \pm 0.098$ | $0.92 \pm 0.11$ | 79 |
| $Q_{i=2}$ | $0.22 \pm 0.034$ | $0.13 \pm 0.018$ | 85 |
| $Q_{i=3}$ | $0.57 \pm 0.072$ | $0.65 \pm 0.084$ | 62 |
| $Q_{i=4}$ | $0.93 \pm 0.12$ | $1.13 \pm 0.15$ | 63 |

At the same time, a high degree of balanced accuracy was achieved—$Ac(V) =$ 79–85% which is higher than the accuracy of azimuthally invariant polarization and Mueller-matrix mapping (see Sect. 3, Tables 3.3, 3.4, 3.7 and 3.8).

## 4.2 Polarization Correlometry of Optically Anisotropic Networks of Biological Layers

### *4.2.1 CDMA Mapping of Biological Layers with Ordered Architectonics*

Figure 4.4 shows the results of an experimental study of the coordinate distributions of CDMA (left side) of a histological section of skeletal muscle tissue, as well as a histogram of the distribution of random values of the CDMA modulus (right side).

It can be seen from the obtained data that the manifestations of optical anisotropy of the birefringent fibrillar network of the histological section of the myocardium spatially ordered along the directions of the optical axes differ significantly in the values of the statistical moments of the first to fourth orders characterizing the distribution of the values of the CDMA modulus from the similar results of polarization-correlation microscopy of this biological layer (see Figs. 4.1 and 4.2).

The main reason for this, in our opinion, is the other "information content" of this polarization-correlation method, which allows direct estimation of the coordinate structure of an optically anisotropic spatially ordered network of myosin fibrils. Based on this, the histograms of the distributions $N(W)$ of random values of the CDMA modulus are even more asymmetric with a fairly small dispersion and a clearly defined extremum of the dependence (see Fig. 4.1, the right side and Fig. 4.4, the right side).

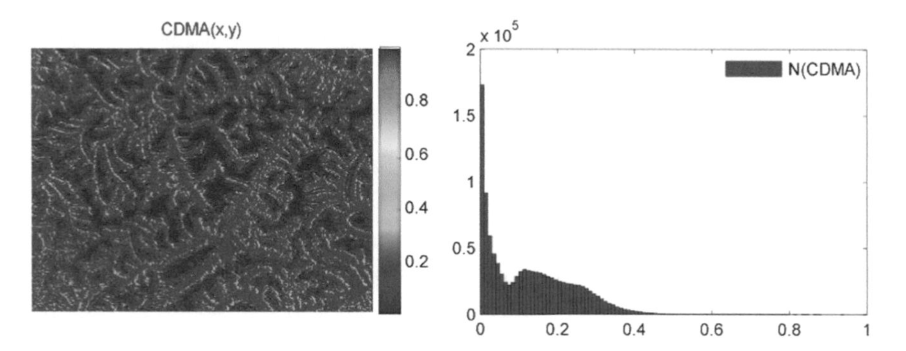


**Fig. 4.4** CDMA map (left side) and a histogram (right side) of the distribution of values of the CDMA modulus of the histological section of the skeletal muscle

**Table 4.5** Statistical moments $Q_{i=1;2;3;4}$ that characterize the coordinate distribution of the values of the CDMA modulus of the histological section of the skeletal muscle

| $Q_{i=1;2;3;4}$ | CDMA |
|---|---|
| $Q_{i=1}$ | 0.08 |
| $Q_{i=2}$ | 0.12 |
| $Q_{i=3}$ | 1.14 |
| $Q_{i=4}$ | 1.68 |

Quantitatively, the results of a statistical analysis of the coordinate distributions of the values of the CDMA modulus of optically thin histological sections of the skeletal muscle are shown in Table 4.5.

A comparative analysis of the magnitude of the set of statistical moments of the first to fourth orders that characterize the coordinate distributions of the values of the modulus CDMP and CDMA polycrystalline networks of the skeletal muscle layer was found (see Tables 4.2 and 4.5):

- Various types of polarization-correlation distributions characterizing the structural heterogeneity of the azimuth and polarization ellipticity at the points of microscopic images and the corresponding orientations of the optical axes and birefringence are characterized by an individual set of statistical moments $Q_{i=1;2;3;4}$.
- The following relationships between the values of statistical parameters:

$$\begin{aligned} Q_1(W) &\prec\prec Q_1(V); \\ Q_2(W) &\approx Q_2(V); \\ Q_3(W) &\succ Q_3(V); \\ Q_4(W) &\succ Q_4(V). \end{aligned} \tag{4.3}$$

- Highest-order statistical moments $Q_{i=3;4}(W)$, which characterize the skewness and sharpness of the peak of the histograms of the distribution of the values of the CDMA modulus of such objects, turned out to be the most sensitive in magnitude to the manifestations of birefringence of spatially ordered polycrystalline networks of skeletal muscle.

### 4.2.2 CDMA Mapping of Biological Layers with Disordered Architectonics

Figure 4.5 shows the results of an experimental study of the coordinate distributions of CDMA (right side) of a histological brain section (tissue with a birefringent network disordered in the directions of the optical axes), as well as histograms of the distribution of random values of the CDMA modulus, characterizing the coordinate consistency of optical anisotropy parameters in the plane of the biological layer (left side).

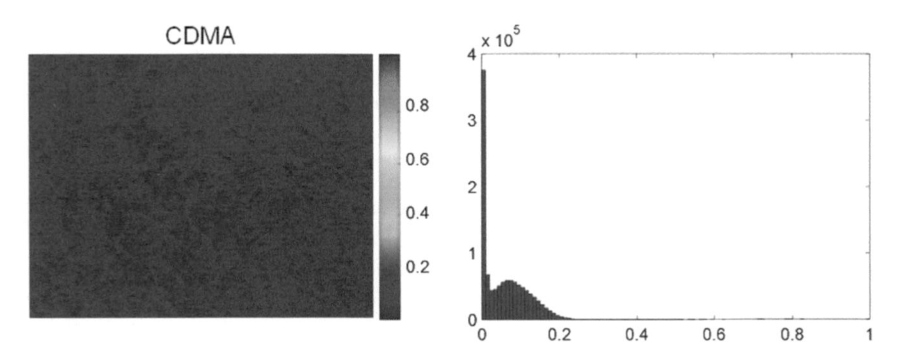


**Fig. 4.5** CDMA map (left side) and a histogram (right side) of the distribution of CDMA modulus values in a microscopic image of a histological section of the brain

Analysis of the obtained data shows that the manifestations of the optical anisotropy of the polycrystalline network of the histological section of brain tissue significantly differ from the similar results of polarization-correlation microscopy of the histological section of the skeletal muscle (see Figs. 4.4 and 4.5).

From a physical point of view, the main factor in such changes is the difference in the spatial-geometric structure of fibrillar networks, which determines the specific distribution of the directions of the optical axes. As a result, a larger spectrum of such directions is formed for a histological section of brain tissue. This, in turn, leads to a wider range of changes in the values of the CDMA modulus in the plane of the biological layer. Based on this, the histograms of the distributions $N(W)$ of random values of the CDMA modulus are less asymmetric with a greater dispersion of the dependence (see Fig. 4.4, the right side and Fig. 4.5, the right side).

Quantitatively, the results of a statistical analysis of the coordinate distributions of the values of the CDMA modulus of the brain tissue layer are given in Table 4.6.

A comparative analysis of the aggregate of statistical parameters characterizing the coordinate distribution of the values of the CDMA modulus of polycrystalline networks of biological layers of both types ($W$ is the skeletal muscle, $W^*$ is the brain tissue) revealed the following relationships:

**Table 4.6** Statistical moments $Q_{i=1;2;3;4}$ that characterize the coordinate distribution of the values of the CDMA modulus of the microscopic image of a histological section of the brain

| $Q_{i=1;2;3;4}$ | CDMA |
|---|---|
| $Q_{i=1}$ | 0.06 |
| $Q_{i=2}$ | 0.19 |
| $Q_{i=3}$ | 0.85 |
| $Q_{i=4}$ | 1.23 |

$$\begin{aligned} Q_1(W) &\prec Q_1(W^*) \\ Q_2(W) &\prec Q_2(W^*) \\ Q_3(W) &\succ Q_3(W^*) \\ Q_4(W) &\succ Q_4(W^*). \end{aligned} \quad (4.4)$$

The obtained results and certain regularities in the formation of CDMA cataracts of networks of spatially ordered and disordered along the directions of the optical axes networks of biological crystals were the basis for the development of a polarization-correlation microscopy of biological layers in differentiating changes in optical anisotropy caused by necrotic changes in the myocardium.

### *4.2.3 Diagnostic Features of CDMA-Mapping*

This section contains the results of studies on the possibility of polarization-correlation differentiation of changes in birefringence of myosin myocardial networks due to necrotic changes in the myocardium.

Figure 4.6 shows the polarization-correlation CDMA maps $W(m \times n)$ (left parts) and histograms $N(W)$ (right parts) of the distributions of CDMA modulus values in the plane of histological sections of the myocardium that deceased due to mechanical asphyxiation (upper line) and heart attack (lower line).

Quantitative differences between polarization-correlation maps characterize the values of statistical moments of the first–fourth orders, which are given in Table 4.7.

From the analysis of statistical moments of the first–fourth orders $Q_{i=1;2;3;4}$, which characterize the coordinate distributions of the values of the modulus of the polarization-correlation parameter CDMA of histological sections of the myocardium of both types, the following main differences are revealed.

First, for both causes of death, there is a significant difference between the values of all four statistical moments $Q_{i=1;2;3;4}(W)$ and $Q_{i=1;2;3;4}(W^*)$.

Second—for necrotic changes caused by a heart attack, there is a decrease in average ($Q_1(W) \downarrow$) and dispersion ($Q_2(W) \downarrow$).

Third, for the case of mechanical asphyxiation, the values of the statistical moments increase $Q_3(W^*) \uparrow$ and $Q_4(W^*) \uparrow$.

The revealed scenarios can be associated with a different effect of necrotic changes on the birefringence of myosin fibrillar networks. For the samples deceased due to a heart attack, there is a morphological destruction of the myocardial fibrillar network. Optically, this is reflected in a decrease in birefringence. As a result, the probability of formation of CDMA modulus values other than extreme decreases $W = 0$. As a result, the dispersion decreases and, conversely, the average, skewness, and kurtosis values increase, characterizing the distribution of this correlation parameter of the histological section of the necrotic altered myocardium.

The expansion of the range of variation of random values of the CDMA modulus in the plane of the birefringent network of the histological section of the myocardium

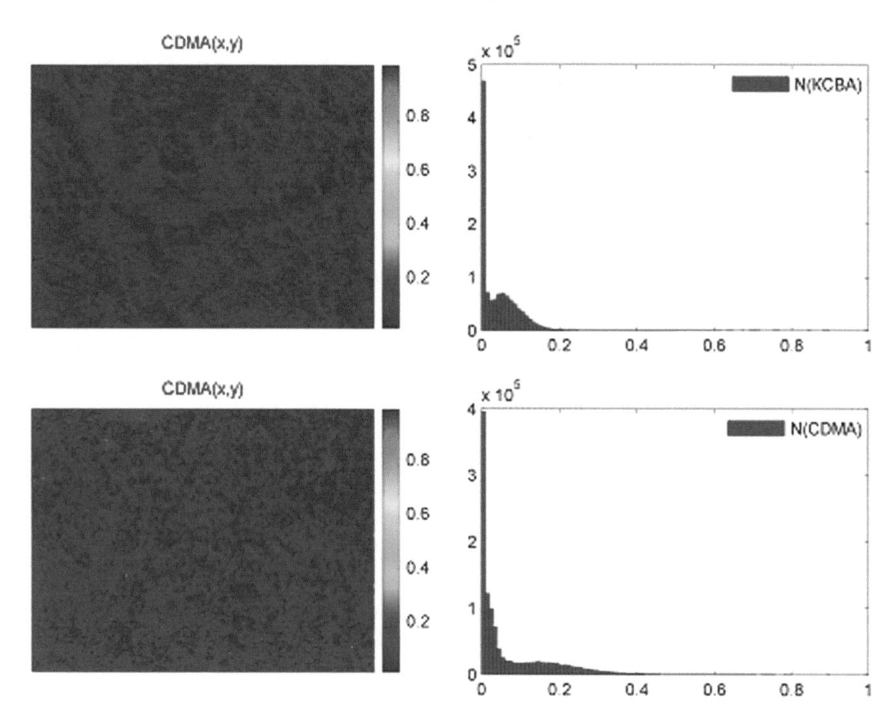


**Fig. 4.6** CDMA maps (left parts) and histograms (right parts) of the distribution of CDMA modulus values in microscopic images of histological sections of the myocardium that deceased from mechanical asphyxiation (top line) and heart attack (bottom line)

**Table 4.7** Statistical moments $Q_{i=1;2;3;4}$ that characterize the distribution of the values of the CDMA modulus of histological sections of the myocardium that deceased due to mechanical asphyxiation ($W$) and heart attack ($W^*$)

| $Q_{i=1;2;3;4}$ | $W$ | $W^*$ | $Ac(\%)$ |
|---|---|---|---|
| $Q_{i=1}$ | $0.12 \pm 0.017$ | $0.07 \pm 0.009$ | 83 |
| $Q_{i=2}$ | $0.19 \pm 0.025$ | $0.11 \pm 0.018$ | 84 |
| $Q_{i=3}$ | $1.09 \pm 0.16$ | $1.48 \pm 0.021$ | 90 |
| $Q_{i=4}$ | $1.63 \pm 0.23$ | $2.12 \pm 0.29$ | 86 |

that deceased due to mechanical asphyxiation is associated with a high level of birefringence of the myosin network of this sample—$Q_1(W) \downarrow$; $Q_2(W) \uparrow$; $Q_3(W) \downarrow$; $Q_4(W) \downarrow$.

The results of the comparative analysis of the averaged (within the respective groups of samples) values of the statistical moments characterizing the two-dimensional distribution of the values of the CDMA modulus are given in Table 4.7.

The data in Table 4.7 allowed us to establish criteria for intergroup differences in the values of statistical moments of the first–fourth orders, which characterize the polarization-correlation CDMA maps $W(m \times n)$ of histological sections of the myocardium of both types

$$W(m \times n) \Leftrightarrow \{\Delta M_1 = 1,22;\ \Delta M_2 = 1,26;\ \Delta M_3 = 1,34;\ \Delta M_4 = 1,36.$$

The results revealed the sensitivity of all four statistical moments $Q_{i=1;2;3;4}$ to necrotic changes in the optical anisotropy of polycrystalline myosin networks of the myocardium. At the same time, the highest level of balanced accuracy was achieved—$Ac(W) = 80$–$87\%$ which is higher (10–20%) than the accuracy of azimuthally invariant polarization and Mueller-matrix mapping using wavelet analysis and spatial-frequency filtering.

## References

1. V. Tuchin, L. Wang, D. Zimnjakov, *Optical Polarization in Biomedical Applications* (Springer, New York, USA, 2006)
2. R. Chipman, in *Polarimetry*, ed. by M. Bass. Handbook of Optics: Vol I—Geometrical and Physical Optics, Polarized Light, Components and Instruments (McGraw-Hill Professional, New York, 2010), pp. 22.1–22.37
3. N. Ghosh, M. Wood, A. Vitkin, in *Polarized Light Assessment of Complex Turbid Media Such as Biological Tissues Via Mueller Matrix Decomposition*, ed. by V. Tuchin. Handbook of Photonics for Biomedical Science (CRC Press, Taylor & Francis Group, London, 2010), pp. 253–282
4. S. Jacques, Polarized light Imaging of Biological Tissues, in *Handbook of Biomedical Optics*, ed. by D. Boas, C. Pitris, N. Ramanujam (CRC Press, Boca Raton, London, New York, 2011), pp. 649–669
5. N. Ghosh, Tissue polarimetry: concepts, challenges, applications, and outlook. J. Biomed. Opt. **16**(11), 110801 (2011)
6. M. Swami, H. Patel, P. Gupta, Conversion of 3 × 3 Mueller matrix to 4 × 4 Mueller matrix for non-depolarizing samples. Opt. Commun. **286**, 18–22 (2013)
7. D. Layden, N. Ghosh, A. Vitkin, in *Quantitative Polarimetry for Tissue Characterization and Diagnosis*, ed. by R. Wang. Advanced Biophotonics: Tissue Optical Sectioning (CRC Press, Taylor & Francis Group, Boca Raton, London, New York, 2013), pp. 73–108
8. T. Vo-Dinh, *Biomedical Photonics Handbook*, 3 vol. set, 2nd edn. (CRC Press, Boca Raton, 2014)
9. A. Vitkin, N. Ghosh, A. Martino, Tissue polarimetry, in *Photonics: Scientific Foundations, Technology and Applications*, 4th edn., ed. by D. Andrews (Wiley, Hoboken, New Jersey, 2015), pp. 239–321
10. V. Tuchin, *Tissue optics: Light Scattering Methods and Instruments for Medical Diagnosis*, 2nd edn. (SPIE Press, Bellingham, Washington, USA, 2007)
11. W. Bickel, W. Bailey, Stokes vectors, Mueller matrices, and polarized scattered light. Am. J. Phys. **53**(5), 468–478 (1985)
12. A. Doronin, C. Macdonald, I. Meglinski, Propagation of coherent polarized light in turbid highly scattering medium. J. Biomed. Opt. **19**(2), 025005 (2014)
13. A. Doronin, A. Radosevich, V. Backman, I. Meglinski, Two electric field Monte Carlo models of coherent backscattering of polarized light. J. Opt. Soc. America A **31**(11), 2394 (2014)
14. A. Ushenko, V. Pishak, in *Laser Polarimetry of Biological Tissue: Principles and Applications*, ed. by V. Tuchin. Handbook of Coherent-Domain Optical Methods: Biomedical Diagnostics (Environmental and Material Science, 2004), pp. 93–138

15. E. Wolf, Unified theory of coherence and polarization of random electromagnetic beams. Phys. Lett. A **312**, 263–267 (2003)
16. J. Tervo, T. Setala, A. Friberg, Degree of coherence for electromagnetic. Opt. Express **11**, 1137–1143 (2003)
17. J.M. Movilla, G. Piquero, R. Martínez-Herrero, P.M. Mejías, Parametric characterization of non-uniformly polarized. Opt. Commun. **149**, 230–234 (1998)
18. J. Ellis, A. Dogariu, Complex degree of mutual polarization. Opt. Lett. **29**, 536–538 (2004)
19. C. Mujat, A. Dogariu, Statistics of partially coherent beams: a numerical analysis. J. Opt. Soc. Am. A **21**(6), 1000–1003 (2004)
20. F. Gori, Matrix treatment for partially polarized, partially coherent beams. Opt. Lett. **23**, 241–243 (1998)
21. E. Wolf, Significance and measurability of the phase of a spatially coherent optical field. Opt. Lett. **28**, 5–6 (2003)
22. M. Mujat, A. Dogariu, Polarimetric and spectral changes in random electromagnetic fields. Opt. Lett. **28**, 2153–2155 (2003)
23. J. Ellis, A. Dogariu, S. Ponomarenko, E. Wolf, Interferometric measurement of the degree of polarization and control of the contrast of intensity fluctuations. Opt. Lett. **29**, 1536–1538 (2004)

# Chapter 5
# Multifunctional Stokes Correlometry of Biological Layers

## 5.1 Wavelet Analysis of CDMP Maps of Microscopic Images of Biological Tissues

### 5.1.1 *Biological Tissues with Ordered Architectonics*

Figure 5.1 and Table 5.1 below present the results of the wavelet analysis of the coordinate distributions of the values of the CDMP modulus of microscopic images of a histological section of the skeletal muscle.

The analysis of the obtained data revealed a large degree of polarization uniformity of the microscopic image of the optically anisotropic spatially ordered fibrillar network of the skeletal muscle at small geometric scales, which are formed by optically active chains of myosin molecules in comparison with large window scales of the MHAT function. This fact is indicated by the different "structures" of the distribution of the values of the wavelet coefficients $W_{a,b}\{V(x, y)\}$ and their different-scale anharmonic distributions $C_{a,b}$. As can be seen, at the small size window of the wavelet function, a rather harmonious (with a small modulation depth) modulation of the amplitude of the wavelet-coefficients $W_{a=\text{min},b}$ of the coordinate distribution of the values of the CDMP modulus of the points of the microscopic image is formed. The opposite picture—the formation of anharmonic distributions (with a large modulation depth) of the amplitude of the wavelet-coefficients $W_{a=\text{max},b}$—takes place for the large-scale MHAT window—a function that distinguishes the manifestations of linear birefringence of the fibrillar network.

From a physical point of view, the differences revealed can be explained by the following considerations. The first is a comparison of the polarization states at the "nearby" (1 pix $\leftrightarrow$ 2 $\mu$m) points of the microscopic image of local microfibrils which forms extremely large CDMP values ($V \rightarrow 1$—Table 4.1) due to small differences between the azimuth and polarization ellipticity. In accordance with this, the small-scale section of the wavelet-coefficient map is a dependence that is formed by a



I. Meglinski et al., *Shedding the Polarized Light on Biological Tissues*, SpringerBriefs in Applied Sciences and Technology,
https://doi.org/10.1007/978-981-10-4047-4_5

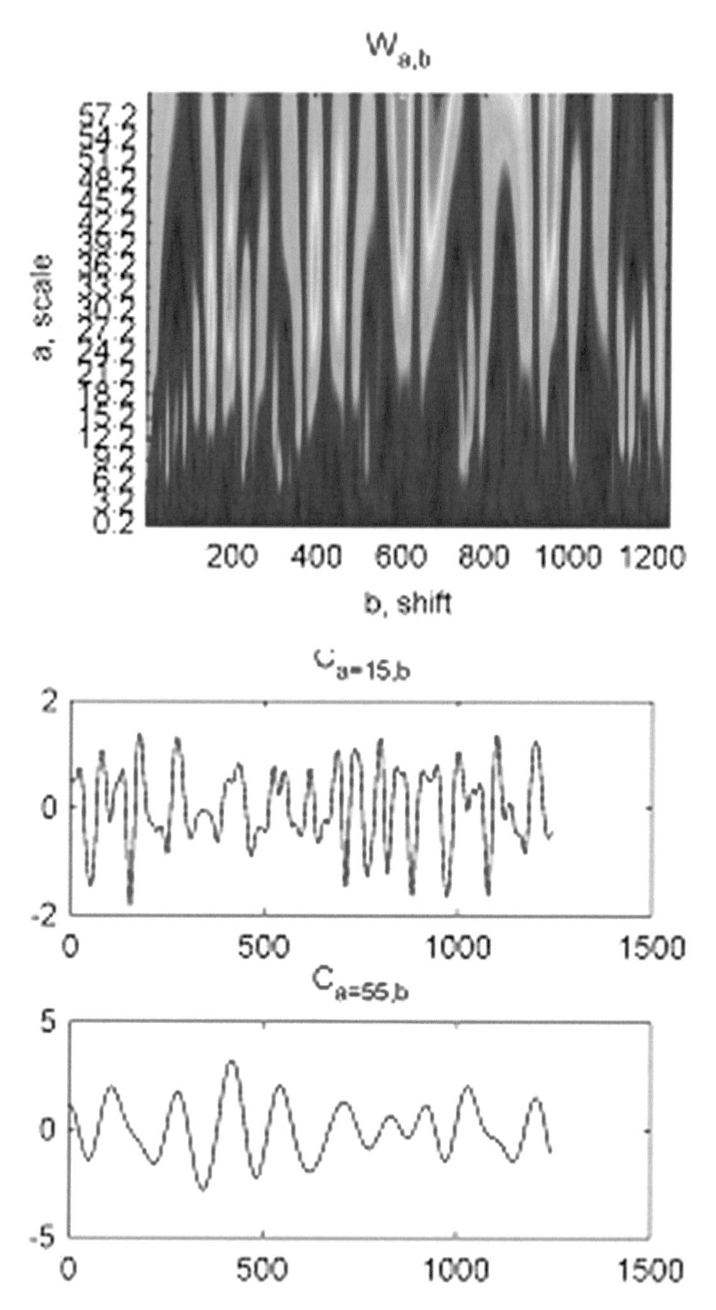


**Fig. 5.1** Wavelet-coefficients and their multiscale cross sections of the distribution of the values of the CDMP modulus of the microscopic image of a histological section of the skeletal muscle

**Table 5.1** Statistical moments $Q_{i=1;2;3;4}$ that characterize the distribution of the values of the CDMP modulus of the microscopic image of a histological section of the skeletal muscle

| Parameters | CDMP | CDMP | CDMP |
|---|---|---|---|
| $a$ | 5 | 25 | 80 |
| $Q_{i=1}$ | 0.012 | 0.025 | 0.058 |
| $Q_{i=2}$ | 0.09 | 0.14 | 0.27 |
| $Q_{i=3}$ | 0.42 | 0.87 | 1.42 |
| $Q_{i=4}$ | 0.74 | 0.53 | 0.34 |

superposition of a small number of harmonics. Each such harmonic corresponds to a certain stable state of polarization. An increase in correlation analysis leads to the formation of a wide range of changes in the CDMP modulus due to the fact that the polarization states at the "far-spaced" points of the microscopic image are quite different due to variations in the directions of the optical axes and the birefringence of myosin fibrils. Accordingly, the depth of modulation of dependencies increases $C_{a=\max,b}$.

Quantitatively, the possibility of an objective polarization-correlation assessment of birefringence manifestations at different scales of the dimensions of a spatially structured fibrillar network of skeletal muscle by wavelet analysis of the coordinate distributions of the values of the CDMP modulus of a microscopic image is illustrated in Table 5.1.

The analysis of the obtained data revealed that all statistical moments of the first–fourth orders characterizing the distribution of the values of wavelet-coefficients $W_{a,b}\{V(x, y)\}$ are nonzero.

The following dynamics of the statistical moments of the third–fourth orders of magnitude characterizing the distribution of the amplitudes of the wavelet-coefficients $W_{a,b}$ from the region of small $\left(W_{a=\min,b}\{V(x, y)\}\right)$ to large $\left(W_{a=\max,b}\{V(x, y)\}\right)$ window scales of the wavelet function was revealed: $\left\{\begin{array}{c} a \Uparrow \\ Q_1\left(C_{a,b}\right) \Uparrow \\ Q_2\left(C_{a,b}\right) \Uparrow \\ Q_3\left(C_{a,b}\right) \Uparrow \\ Q_4\left(C_{a,b}\right) \Downarrow \end{array}\right\}$, before $\Uparrow$—increase in the value; $\Downarrow$—decrease in value.

### 5.1.2 *Biological Tissues with Disordered Architectonics*

Figure 5.2 shows a series of dependences of two-dimensional and linear dependences of wavelet-coefficients characterizing the results of a scale-selective analysis of the coordinate distribution of the values of the CDMP modulus of a microscopic image of a histological section of the brain.

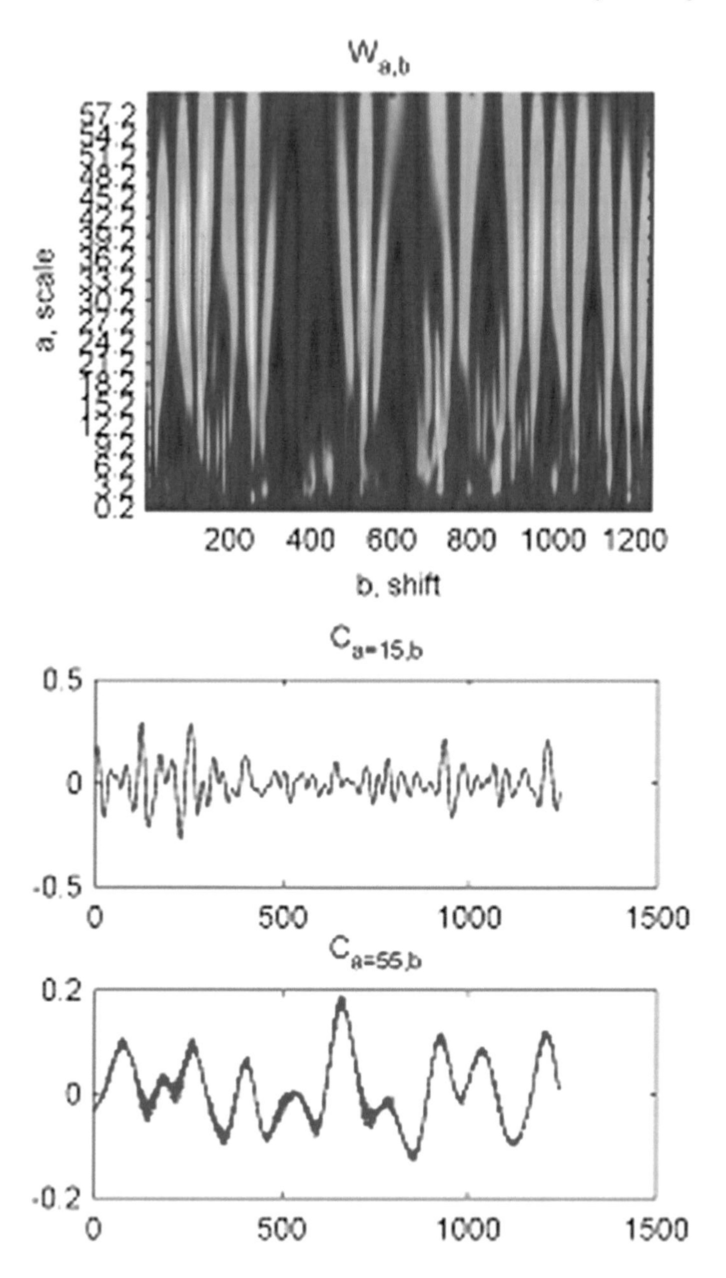


**Fig. 5.2** Wavelet-coefficients and their multiscale cross sections of the distribution of the values of the CDMP modulus of a microscopic image of a histological section of the brain

**Table 5.2** Statistical moments $Q_{i=1;2;3;4}$ that characterize the distribution of values of the CDMP modulus of a microscopic image of a histological section of the brain

| Parameters | CDMP | CDMP | CDMP |
|---|---|---|---|
| $a$ | 5 | 25 | 80 |
| $Q_{i=1}$ | 0.019 | 0.033 | 0.065 |
| $Q_{i=2}$ | 0.12 | 0.18 | 0.32 |
| $Q_{i=3}$ | 0.74 | 1.27 | 1.84 |
| $Q_{i=4}$ | 0.92 | 0.75 | 0.53 |

A comparative analysis of the results of wavelet scanning data of two-dimensional distributions of polarization-correlation maps $V(x, y)$ of a microscopic image of a histological section of the brain and a similar study of a birefringent fibrillar network of a microscopic image of a histological section of a skeletal muscle revealed the formation of a higher level of distribution modulation $C_{a,b}$ at all scales $a$ of the window of the MHAT function.

This is related to the fact that a network of optically anisotropic protein fibers of brain tissue disordered in the directions of the optical axes "forms" a larger coordinate spectrum of the azimuth and polarization ellipticity. As a result of this, the amplitudes of the wavelet-coefficients $W_{a,b}$ at different scales of the window $a$ of the wavelet function $\Omega_{a,b}$, which scans the polarization-correlation maps $\{V(x, y)\}$, vary in a larger range. In accordance with this, other values of statistical moments of the first to fourth orders are formed that characterize such distributions—Table 5.2.

The data in Table 5.2 indicates a higher level (compared with the previous results—Table 5.1) of the polarization heterogeneity of the microscopic image of a histological section of brain tissue at all scales of analysis of its correlation structure. Quantitatively, this is indicated by the fact that the differences between the values of all statistical moments of the first to fourth orders, which characterize different-scale sections $C_{a=15,b}$ and $C_{a55,b}$ two-dimensional distributions of wavelet-coefficients $W_{a,b}\{V(x, y)\}$, are 2–3 times smaller than those shown in Table 5.1.

Thus, we have established the sensitivity at different scales of wavelet scanning of two-dimensional distributions of the values of the CDMP modulus of the points of microscopic images of polycrystalline networks of biological tissues with spatially ordered and disordered directions of the optical axes. The obtained information was the basis for the development of the method of polarization-correlation CDMP microscopy of weak changes in optical anisotropy due to necrotic changes in the myocardium of the deceased due to various reasons—mechanical asphyxiation and heart attack.

### 5.1.3 Diagnostic Capabilities of CDMP Mapping of Microscopic Images of Biological Tissues

The results of the wavelet analysis of polarization-correlation CDMP maps of microscopic images of the myocardium histological sections deceased due to mechanical asphyxiation and heart attack are shown in Fig. 5.3.

A comparative analysis of the results of the study of wavelet-coefficients $W_{a,b}(V)$ revealed the greatest differences between them at the average ($a = 25$) and large ($a = 80$) scales of the MHAT function. The revealed differences can be attributed to the fact that necrotic changes in the polycrystalline networks of the myocardium deceased due to mechanical asphyxiation occur not at the morphological (large-scale), but at smaller ("concentration") levels. However, in the framework of polarization-correlation analysis, which is sensitive to differences between the polarization states of the points of the microscopic image of such a network,

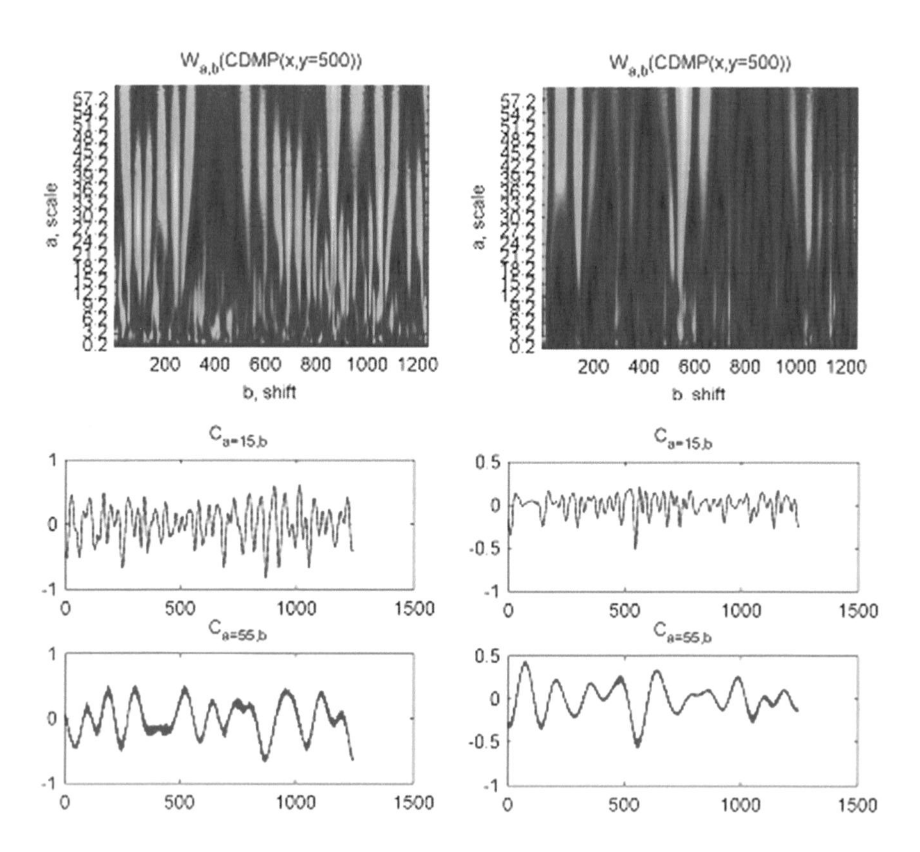


**Fig. 5.3** Wavelet-coefficients and their different-scale cross sections of the distributions of the values of the CDMP modulus of microscopic images of the histological sections of myocardium deceased due to mechanical asphyxiation (left column) and heart attack (right column)

anisotropy changes ($V < 1$) begin to appear on average fibril sizes. Quantitatively, such a process is manifested in the growth of statistical moments of the second and fourth orders $Q_{i=2;3;4}(C_{a,b}(V)) \uparrow$ characterizing the distribution of the values of linear sections $C_{a,b}$ of a two-dimensional wavelet map $W_{a,b}\{V(x, y)\}$ at the average ($a = 25$) and large ($a = 80$) scales of the wavelet function window—Table 5.3.

The following differences (highlighted in italic) between the statistical parameters that characterize the distribution of values of different-scale sections of the values of wavelet maps $W_{a,b}\{V(x, y); V^*(x, y)\}$ of polycrystalline networks of samples of the histological sections of myocardium of both types $\begin{matrix} V(m \times n) \Leftrightarrow a_{midle} \to \Delta Q_2 = 1,45; \Delta Q_3 = 1,29; \Delta Q_4 = 1,52; \\ V(m \times n) \Leftrightarrow a_{\max} \to \Delta Q_2 = 1,53; \Delta Q_3 = 1,43; \Delta Q_4 = 1,41 \end{matrix}$ are established.

The value of the balanced accuracy of the method of mapping the coordinate distributions of the values of the CDMP modulus for microscopic images of the myocardium histological sections deceased due to mechanical asphyxiation and heart attack is shown in Table 5.4.

As can be seen, the scale-selective method for the analysis of polarization-correlation mapping data provides a high level of balanced accuracy ($Ac = 85$–$90\%$) of differentiation of myocardial necrotic changes.

## 5.2 Wavelet Analysis of Biological Tissue CDMA-Maps

### *5.2.1 Biological Tissues with Ordered Architectonics*

Figure 5.4 and Table 5.5 below present the results of a wavelet analysis of the distributions of the values of the modulus of the complex degree of mutual anisotropy of spatially ordered optically anisotropic networks of the histological section of the skeletal muscle.

Comparison of the obtained data with the results of a scale-selective study of the polarization-correlation structure of the microscopic image of the skeletal muscle revealed similar scenarios for its spatially ordered fibrillar network at all geometric scales. At small scales, an insignificant depth of modulation of the amplitude of the wavelet-coefficients $W_{a=\min,b}$ of the coordinate distribution of the values of the CDMA modulus. At medium and large scales—the formation of the distribution of the magnitude of the amplitude of wavelet-coefficients $W_{a=\max,b}$ with a greater modulation depth.

From the physical point of view, the revealed regularities can be explained by small differences between the birefringence parameters at small scales of the geometric dimensions of fibrils. An increase in the scale of correlation analysis leads to the formation of a wide range of changes in the CDMA modulus due to the fact that at the "far-spaced" points of the fibrillar network, there are quite different directions of the optical axes and the magnitude of birefringence.

**Table 5.3** Statistical moments $Q_{i=1;2;3;4}$ that characterize the distribution of the values of the decomposition of wavelet-coefficients of the values of the CDMP modulus of microscopic images of the myocardium histological sections deceased due to mechanical asphyxiation ($V$) and heart attack ($V^*$)

| Parameters | $V$ | $V^*$ | $V$ | $V^*$ | $V$ | $V^*$ |
|---|---|---|---|---|---|---|
| $a$ | 5 | | 25 | | 80 | |
| $Q_{i=1}$ | $0.095 \pm 0.001$ | $0.088 \pm 0.009$ | $0.021 \pm 0.003$ | $0.017 \pm 0.0026$ | $0.043 \pm 0.007$ | $0.035 \pm 0.0046$ |
| $Q_{i=2}$ | $0.11 \pm 0.016$ | $0.095 \pm 0.0013$ | *0.16 ± 0.022* | *0.11 ± 0.018* | *0.23 ± 0.032* | *0.15 ± 0.015* |
| $Q_{i=3}$ | $0.54 \pm 0.077$ | $0.46 \pm 0.059$ | *0.81 ± 0.11* | *0.63 ± 0.088* | *1.27 ± 0.18* | *0.89 ± 0.12* |
| $Q_{i=4}$ | $0.63 \pm 0.084$ | $0.54 \pm 0.054$ | *0.44 ± 0.058* | *0.29 ± 0.037* | *0.31 ± 0.043* | *0.22 ± 0.031* |

**Table 5.4** Balanced accuracy of the method of mapping the coordinate distributions of the values of the CDMP modulus of microscopic images of the myocardium histological sections deceased due to mechanical asphyxiation and heart attack

| $Q_{i=1;2;3;4}$ | $Ac(\%)$ | $Ac(\%)$ |
|---|---|---|
| $a$ | 25 | 80 |
| $Q_{i=1}$ | 63 | 67 |
| $Q_{i=2}$ | 87 | 90 |
| $Q_{i=3}$ | 74 | 88 |
| $Q_{i=4}$ | 90 | 85 |

Quantitatively, the possibilities of an objective description of the manifestations of optical anisotropy at different scales of the size of the myosin network of the histological section of the skeletal muscle by wavelet analysis of the coordinate distributions of the values of the CDMA modulus are illustrated in Table 5.5.

An analysis of the obtained data revealed that, as in the case of a scale-selective correlation analysis of polarization-inhomogeneous images, all statistical moments of the first and fourth orders characterizing the distribution of the values of the wavelet-coefficients $W_{a,b}\{W(x, y)\}$ are nonzero.

The dynamics of the magnitude of the statistical moments of the first and fourth orders characterizing the distribution of the amplitudes of the wavelet-coefficients $W_{a,b}\{W(x, y)\}$ is also similar for different scales of the wavelet function window

$$W_{a,b} \Leftrightarrow \left\{ \begin{array}{c} a \ \Uparrow \\ Q_1(C_{a,b}) \Uparrow \\ Q_2(C_{a,b}) \Uparrow \\ Q_3(C_{a,b}) \Uparrow \\ Q_4(C_{a,b}) \Downarrow \end{array} \right\} \text{: before } \Uparrow\text{—an increase in the value; } \Downarrow\text{—decrease in value.}$$

## 5.2.2 Biological Tissues with Disordered Architectonics

Figure 5.5 and Table 5.6 below present the results of the wavelet analysis of the distributions of the values of the modulus of the complex degree of mutual anisotropy of spatially ordered optically anisotropic networks of the histological brain section.

A comparison of the results of a scale-selective analysis of the distributions of the polarization-correlation CDMA maps $W(x, y)$ of the birefringent fibrillar network of the histological section of the brain and a similar study of the distributions of the parameters of the optical anisotropy of the histological section of the skeletal muscle revealed a large modulation of the distributions $C_{a,b}$ at all scales $a$ of the MHAT function window.

This is due to the fact that, as it is disordered in the directions of laying, the network of protein fibers of brain tissue has a larger coordinate spectrum of orientations of the optical axes and the magnitude of birefringence. In accordance with this, other values

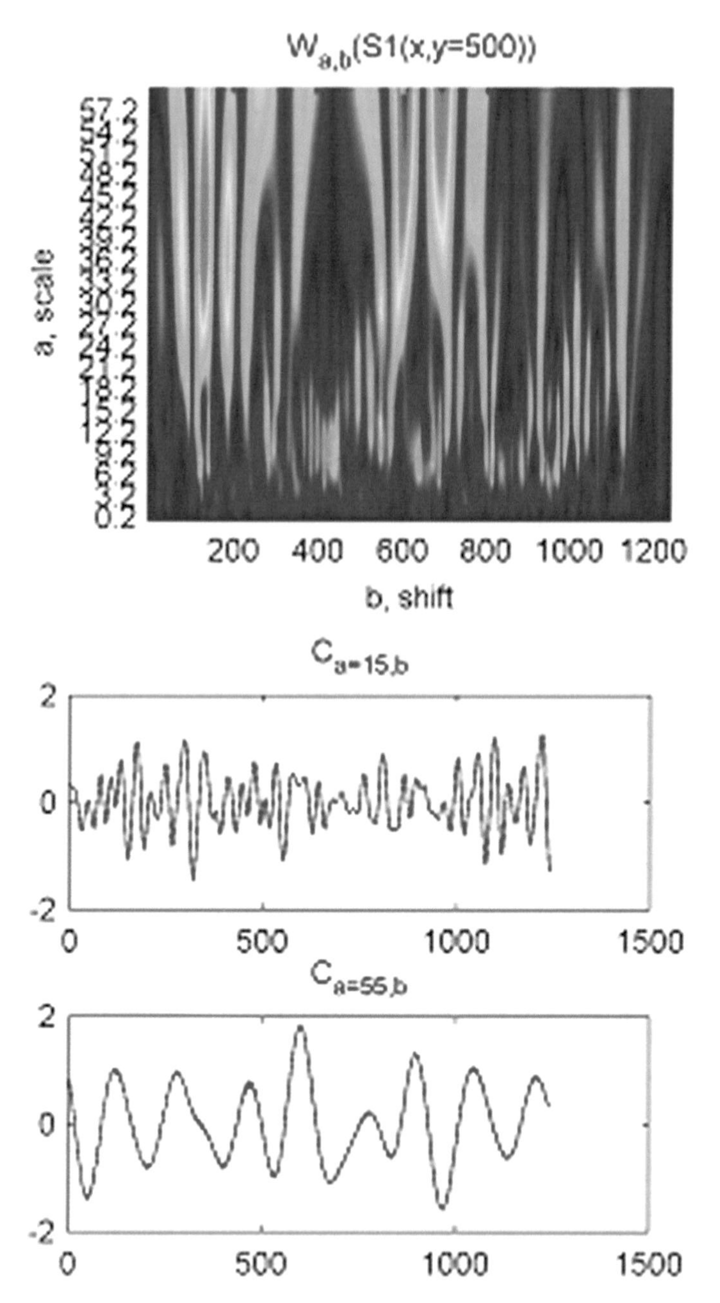


**Fig. 5.4** Wavelet-coefficients and their multiscale cross sections of the distribution of the values of the CDMA modulus of the histological section of the skeletal muscle

**Table 5.5** Statistical moments $Q_{i=1;2;3;4}$ that characterize the distribution of the values of the CDMA modulus of the microscopic image of a histological section of the skeletal muscle

| Parameters | CDMA | CDMA | CDMA |
|---|---|---|---|
| $a$ | 5 | 25 | 80 |
| $Q_{i=1}$ | 0.033 | 0.042 | 0.063 |
| $Q_{i=2}$ | 0.11 | 0.14 | 0.21 |
| $Q_{i=3}$ | 0.64 | 1.08 | 1.74 |
| $Q_{i=4}$ | 0.97 | 0.75 | 0.53 |

of the statistical moments of the first–fourth orders are formed that characterize such distributions—Table 5.6.

The data presented in Table 5.6 indicates a higher level (compared with previous results—Table 5.5) of the coordinate heterogeneity of the distributions of the values of the birefringence parameters of the histological section of the brain tissue at all scales of the analysis of manifestations of optical anisotropy. Quantitatively, this is indicated by the fact that the differences between the values of all statistical moments of the first–fourth orders, which characterize different-scale sections $C_{a=15,b}$ and $C_{a55,b}$ two-dimensional distributions of wavelet-coefficients $W_{a,b}\{V(x, y)\}$, are 1.5–2 times smaller than those shown in Table 5.5.

The obtained information was the basis for the development of a method of CDMA-mapping of the differentiation of changes in optical anisotropy due to necrotic changes in the myocardium of the deceased due to various reasons—mechanical asphyxiation and heart attack.

### 5.2.3 *Diagnostic Capabilities of CDMA Mapping of Biological Tissue Images*

The results of the wavelet analysis of polarization-correlation CDMA cards of optically anisotropic birefringent fibrillar networks of the myocardium histological sections dead due to mechanical asphyxiation and heart attack are shown in Fig. 5.6.

A comparative analysis of the results of studying the distributions of wavelet-coefficients $W_{a,b}(W)$ with the data of a scale-selective analysis of coordinate distributions of CDMP modulus values (see Table 5.4) also found the greatest differences between them at the level of average ($a = 25$) and large ($a = 80$) scales of the MHAT function.

The statistical moments of the second–fourth orders of magnitude $(Q_{i=2;3;4}(C_{a,b}(W)) \uparrow)$, which characterize the distributions of the values of linear sections $C_{a,b}$ at the average ($a = 25$) and large ($a = 80$) scales of the wavelet function window, turned out to be the most sensitive to changes in optical anisotropy due to necrotic myocardial processes—Table 5.7.

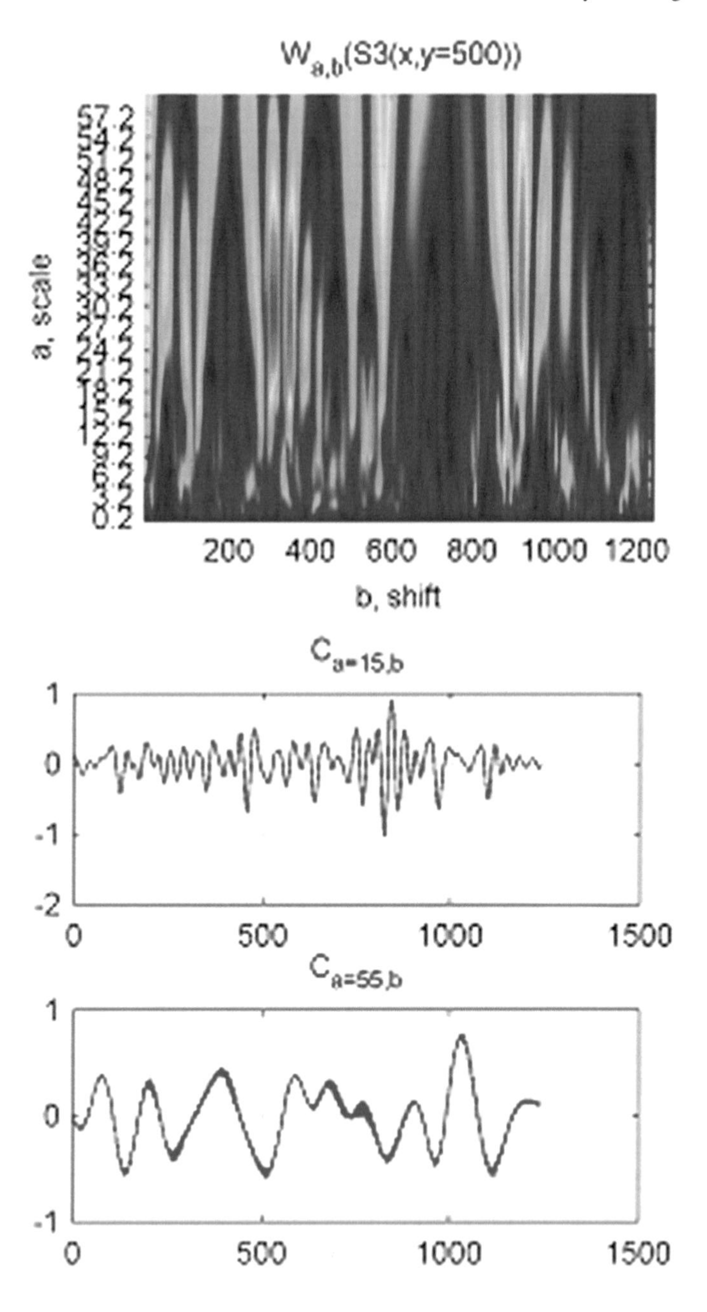


**Fig. 5.5** Wavelet-coefficients and their multiscale cross sections of the distribution of values of the CDMA modulus of a histological brain section

**Table 5.6** Statistical moments $Q_{i=1;2;3;4}$ that characterize the distribution of values of the CDMA modulus of a microscopic image of a histological section of the brain

| Parameters | CDMA | CDMA | CDMA |
|---|---|---|---|
| $a$ | 5 | 25 | 80 |
| $Q_{i=1}$ | 0.029 | 0.038 | 0.051 |
| $Q_{i=2}$ | 0.16 | 0.13 | 0.11 |
| $Q_{i=3}$ | 0.62 | 0.84 | 1.14 |
| $Q_{i=4}$ | 0.54 | 0.32 | 0.21 |

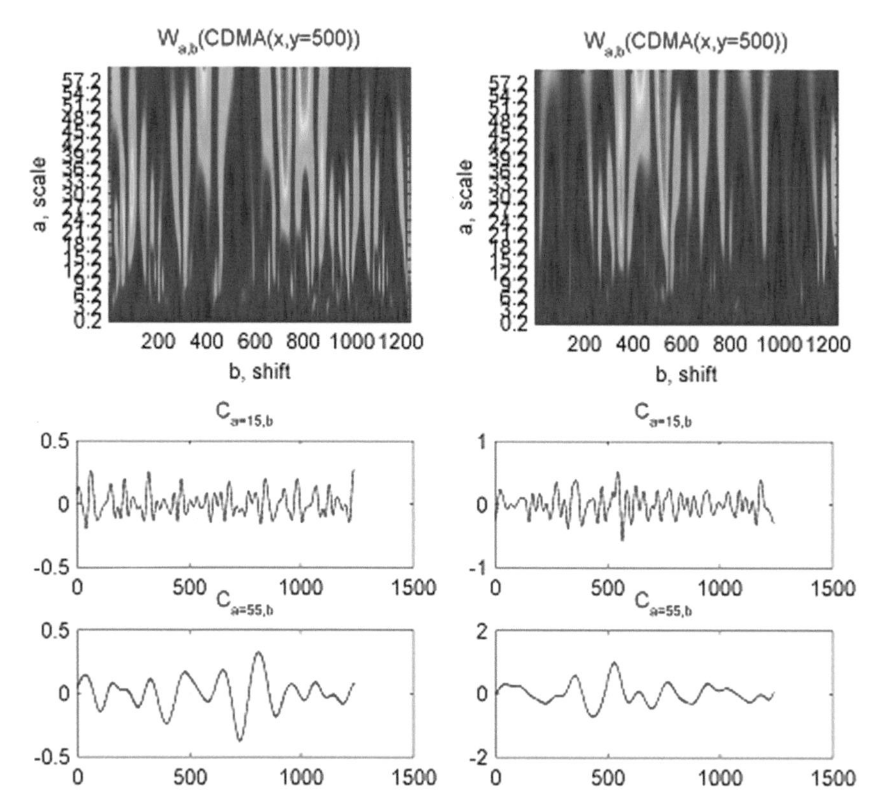


**Fig. 5.6** Wavelet-coefficients and their different-scale cross sections of the distributions of the values of the CDMA modulus of the myocardium histological sections deceased due to mechanical asphyxiation (left column) and heart attack (right column)

The following differences (highlighted in italic) between the treatment-sensitive statistical moments of the second–fourth orders of magnitude characterizing the distribution $C_{a=25;80;b}$ of wavelet-coefficients $W_{a,b}\{W(x,y);\, W^{*}(x,y)\}$ of the CDMA maps of polycrystalline networks

**Table 5.7** Statistical moments $Q_{i=1;2;3;4}$ that characterize the distribution of the values of the CDMA modulus of the myocardium histological sections deceased due to mechanical asphyxiation ($W$) and heart attack ($W^*$)

| Параметри | $W$ | $W^*$ | $W$ | $W^*$ | $W$ | $W^*$ |
|---|---|---|---|---|---|---|
| $a$ | 5 | | 25 | | 80 | |
| $Q_{i=1}$ | $0.029 \pm 0.004$ | $0.022 \pm 0.0034$ | $0.038 \pm 0.005$ | $0.033 \pm 0.0047$ | $0.056 \pm 0.008$ | $0.051 \pm 0.0078$ |
| $Q_{i=2}$ | $0.19 \pm 0.025$ | $0.17 \pm 0.025$ | $0.15 \pm 0.022$ | $0.11 \pm 0.018$ | $0.13 \pm 0.018$ | $0.075 \pm 0.0095$ |
| $Q_{i=3}$ | $0.64 \pm 0.086$ | $0.59 \pm 0.074$ | $1.08 \pm 0.17$ | $0.67 \pm 0.11$ | $1.74 \pm 0.22$ | $1.05 \pm 0.16$ |
| $Q_{i=4}$ | $0.97 \pm 0.11$ | $0.87 \pm 0.096$ | $0.75 \pm 0.099$ | $0.46 \pm 0.061$ | $0.53 \pm 0.073$ | $0.32 \pm 0.044$ |

**Table 5.8** Balanced accuracy of the method of mapping the coordinate distributions of the values of the CDMP modulus of microscopic images of the myocardium histological sections deceased due to mechanical asphyxiation and heart attack

| $Q_{i=1;2;3;4}$ | $Ac(\%)$ | $Ac(\%)$ |
|---|---|---|
| $a$ | 25 | 80 |
| $Q_{i=1}$ | 65 | 67 |
| $Q_{i=2}$ | 74 | 91 |
| $Q_{i=3}$ | 82 | 89 |
| $Q_{i=4}$ | 88 | 90 |

of samples of histological sections of myocardium of both types $W(m \times n) \Leftrightarrow a_{midle} \rightarrow \Delta Q_2 = 1,36; \Delta Q_3 = 1,61; \Delta Q_4 = 1,63;$

$W(m \times n) \Leftrightarrow a_{\max} \rightarrow \Delta Q_2 = 1,7; \Delta Q_3 = 1,65; \Delta Q_4 = 1,66$ .

The value of the balanced accuracy of the method of mapping the coordinate distributions of the values of the CDMP modulus of microscopic images of the myocardium histological sections deceased due to mechanical asphyxiation and heart attack is shown in Table 5.8.

A high degree of balanced accuracy ($Ac = 82$–$91\%$) of the proposed method has been achieved, commensurate with the information content of a similar polarization-correlation method for analyzing polarization-inhomogeneous images of such biological layers.

The obtained highest levels of balanced accuracy open up new possibilities in differentiating the changes in the anisotropy of various biological tissues. However, all the results obtained by the analytical method require the development of direct experimental detection and separation of the manifestations of the optical anisotropy mechanisms of birefringent polycrystalline networks of biological tissues.

## 5.3 Diagnostic Capabilities of the Fourier Analysis of CDMP Maps of Microscopic Images of Biological Tissues

Figures 5.7 and 5.8 show the results of polarization-correlation mapping of optically anisotropic networks of histological sections of the myocardium using spatial-frequency filtering of the corresponding polarization-inhomogeneous microscopic images.

A comparative analysis of the data obtained in the "high-frequency" and "low-frequency" filtering modes on the distributions of the values of the modulus of the CDMP of microscopic images at various scales of the geometric dimensions of optically anisotropic myosin networks revealed the following differences.

For the "high-frequency" component [Fig. 5.10, fragments (1), (2)], there is a higher level of circular birefringence of myosin polypeptide chains of myocardial

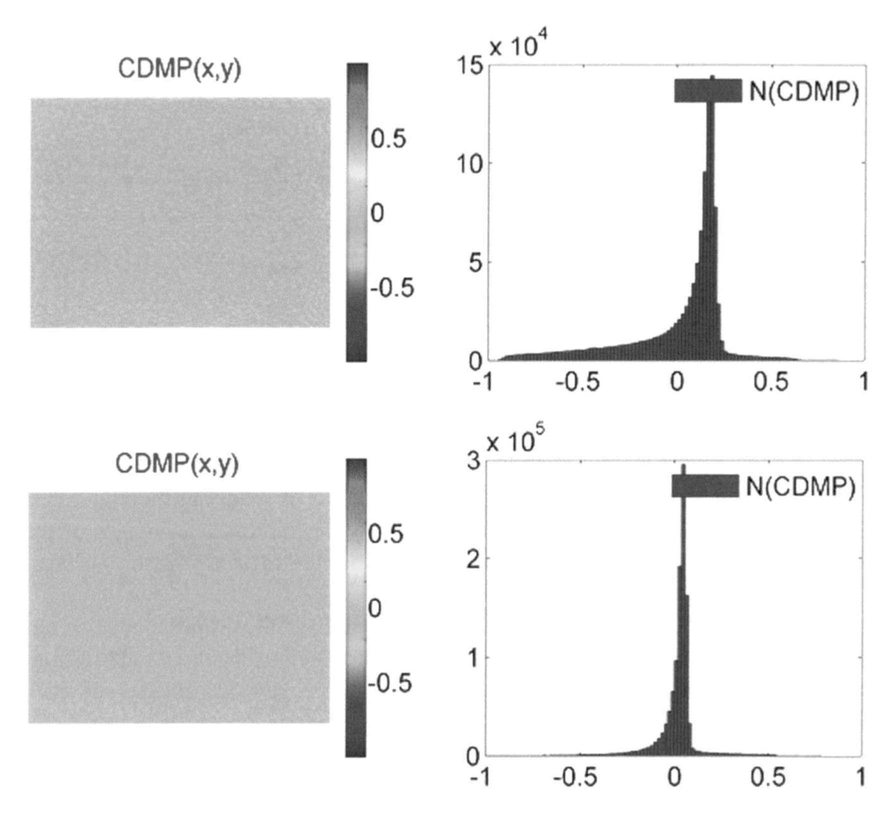


**Fig. 5.7** "High-frequency" (left parts) two-dimensional components and histograms (right parts) of the distributions of the values of the CDMP modulus in microscopic images of the myocardium histological sections deceased due to mechanical asphyxiation (top line) and heart attack (bottom line)

deceased due to mechanical asphyxiation. The corresponding histogram of the distribution of values of the CDMP modulus indicates a higher level of polarization inhomogeneity and is characterized by greater dispersion and skewness see Fig. 5.10, fragments (3), (4)].

For the "low-frequency" component of the distribution map of the values of the CDMP modulus, a decrease in the variance was revealed, which characterizes the corresponding histogram of a microscopic image of a histological section of the myocardium deceased due to heart attack [see Fig. 5.10, fragments (3), (4)] (Table 5.9).

Quantitatively, the distribution of the values of the CDMP modulus in different frequency components of microscopic images of the myocardiumhistological sections deceased due to mechanical asphyxiatoin ($V$) and heart attack ($V^{*}$) illustrates a set of statistical moments $Q_{i=1;2;3;4}$, the values of which are given in Table 5.7.

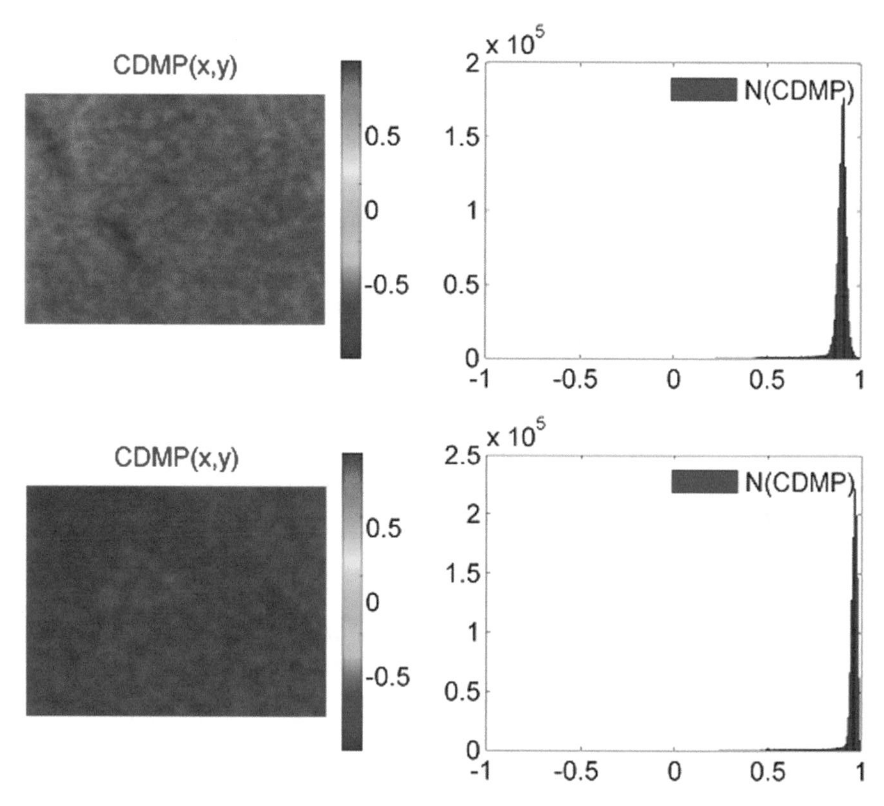


**Fig. 5.8** "Low-frequency" (left parts) two-dimensional components and histograms (right parts) of the distributions of the values of the CDMP modulus in microscopic images of the myocardium histological sections deceased due to mechanical asphyxiation (top line) and heart attack (bottom line)

**Table 5.9** Statistical moments $Q_{i=1;2;3;4}$ that characterize the distribution of the values of the CDMP modulus in different frequency components of microscopic images of the myocardium histological sections deceased due to mechanical asphyxiation ($V$) and heart attack ($V^*$)

| Parameters | $V$ | $V^*$ | $V$ | $V^*$ | $Ac$, % | $Ac$, % |
|---|---|---|---|---|---|---|
| $\nu$ | "High-frequency" component ($\nu^*$) | | "Low-frequency" component ($\nu^{**}$) | | $\nu^*$ | $\nu^{**}$ |
| $Q_{i=1}$ | $0.14 \pm 0.021$ | $0.12 \pm 0.017$ | $0.51 \pm 0.073$ | $0.44 \pm 0.063$ | 63 | 65 |
| $Q_{i=2}$ | $0.11 \pm 0.018$ | $0.09 \pm 0.022$ | $0.27 \pm 0.038$ | $0.16 \pm 0.027$ | 65 | 89 |
| $Q_{i=3}$ | $0.98 \pm 0.11$ | $0.66 \pm 0.087$ | $0.37 \pm 0.051$ | $0.29 \pm 0.038$ | 88 | 72 |
| $Q_{i=4}$ | $1.18 \pm 0.16$ | $1.76 \pm 0.27$ | $0.53 \pm 0.072$ | $0.46 \pm 0.062$ | 90 | 78 |

A comparative analysis of the data revealed the following intergroup differences in the values of statistical moments $Q_{i=1;2;3;4}$ characterizing the distribution of values of the CDMP modulus

$$V(\nu^{*}) \Rightarrow \{\Delta Q_2 = 1,69\};$$
$$V(\nu^{**}) \Rightarrow \{\Delta Q_3 = 1,48;\ \Delta Q_4 = 1,49\}.$$

An analysis of the diagnostic efficiency of the polarization-correlation method for the analysis of microscopic images of histological sections of the myocardium with their spatial-frequency filtering revealed a high level of balanced accuracy $Ac(V^{*}) =$ 88–90%, $Ac(V^{**}) = 90\%$.

## 5.4 Diagnostic Capabilities of the Fourier Analysis of CDMA Cards of Polycrystalline Networks of Biological Tissues

The next step in the application of the spatial-frequency filtering mode was the polarization-correlation mapping of optically anisotropic networks of themy-ocardium histological sections deceased due to mechanical asphyxia (see Fig. 5.9) and heart attack (see Fig. 5.10).

A comparative analysis of the obtained distributions of CDMA modulus values characterizing the coordinate consistency of birefringence parameters at different scales of optically anisotropic protein networks revealed the following differences. For the "high-frequency" component of the CDMA-map $W*$, there is a higher level of circular birefringence of myosin polypeptide chains of the myocardium that deceased due to mechanical asphyxiation (see Fig. 5.9). For a "low-frequency" CDMA card $W^{**}$, there is a decrease in the linear birefringence of myosin networks of the myocardium deceased from a heart attack (see Fig. 5.10).

The quantitative distribution of the values of the modulus of the polarization-correlation parameter CDMA is illustrated by a set of statistical moments $Q_{i=1;2;3;4}$, the values of which are given in Table 5.10.

As can be seen most clearly, the differences between the manifestations of optical anisotropy are observed at the level of large-scale myocardial fibrils with varying degrees of necrotic changes.

The corresponding values of the magnitudes of the statistical moments $Q_{i=1;2;3;4}$ characterizing the distribution of the "different frequency" values of the CDMA modulus differ as follows:

$$W(\nu^{*}) \Rightarrow \{\Delta Q_3 = 1,31;\ \Delta Q_4 = 1,27\};$$
$$W(\nu^{**}) \Rightarrow \{\Delta Q_3 = 1,72;\ \Delta Q_4 = 1,5\}.$$

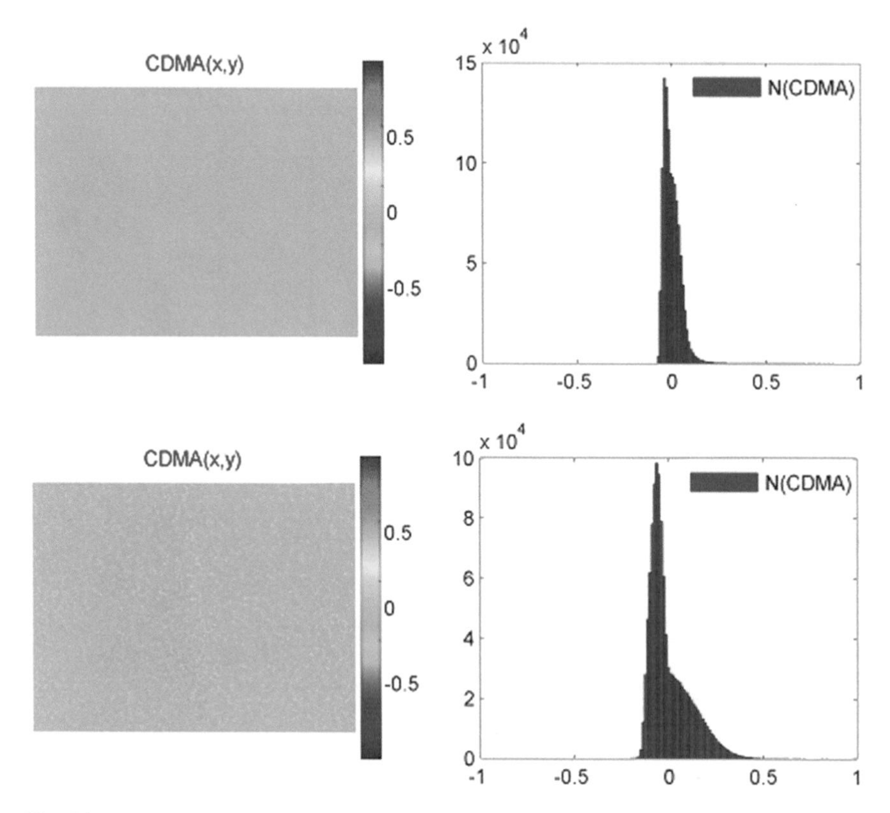


**Fig. 5.9** "High-frequency" (left parts) two-dimensional components and histograms (right parts) of the distributions of the values of the CDMA modulus in microscopic images of the myocardium histological sections deceased due to mechanical asphyxiation (top line) and heart attack (bottom line)

An analysis of the diagnostic effectiveness of this method of polarization-correlation mapping of birefringent networks of histological myocardial sections with spatial-frequency filtering of their polarization-inhomogeneous microscopic images revealed a good level of balanced accuracy $Ac(v^*) = 75–79\%$ for high frequencies according to the criteria of evidence-based medicine, and an excellent level $Ac(v^*) = 86–90\%$ for low frequencies.

## 5.5 Main Results and Conclusions

1. For the first time, a scale-selective wavelet analysis was used to analyze the azimuthally invariant distributions of the azimuth and polarization ellipticities of microscopic Mueller-matrix images of histological sections of biological tissues with structured optically anisotropic fibrillar networks. The diagnostic capabilities of the wavelet analysis of two-dimensional polarization maps

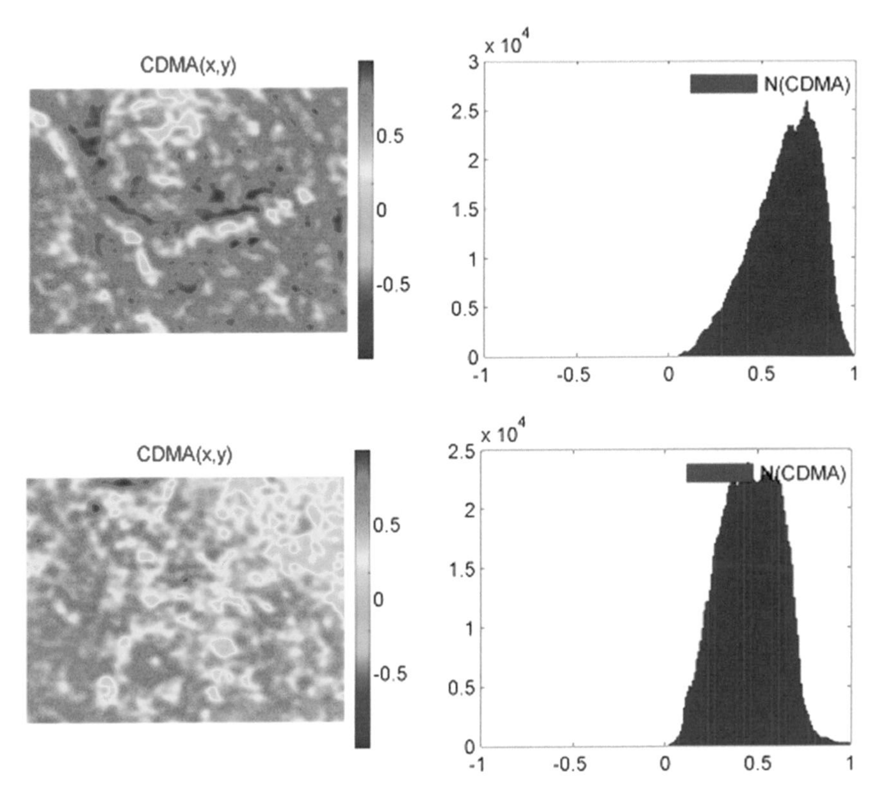


**Fig. 5.10** "Low-frequency" (left parts) two-dimensional components and histograms (right parts) of the distributions of the values of the CDMA modulus in microscopic images of the myocardium histological sections deceased due to mechanical asphyxiation (top line) and heart attack (bottom line)

**Table 5.10** Statistical moments $Q_{i=1;2;3;4}$ that characterize the distribution of the values of the CDMA modulus of the myocardium histological sections deceased due to mechanical asphyxiation ($W$) and heart attack ($W^*$)

| Parameters | $W$ | $W^*$ | $W$ | $W^*$ | $Ac, \%$ | $Ac, \%$ |
|---|---|---|---|---|---|---|
| $\nu$ | "High-frequency" component ($\nu^*$) | | "Low-frequency" component ($\nu^{**}$) | | $\nu^*$ | $\nu^{**}$ |
| $Q_{i=1}$ | $0.105 \pm 0.017$ | $0.09 \pm 0.01$ | $0.41 \pm 0.055$ | $0.35 \pm 0.044$ | 62 | 67 |
| $Q_{i=2}$ | $0.14 \pm 0.019$ | $0.12 \pm 0.016$ | $0.22 \pm 0.031$ | $0.25 \pm 0.033$ | 60 | 63 |
| $Q_{i=3}$ | $0.76 \pm 0.097$ | $0.59 \pm 0.081$ | *$0.38 \pm 0.052$* | *$0.22 \pm 0.028$* | 75 | 90 |
| $Q_{i=4}$ | $1.74 \pm 2.1$ | $1.41 \pm 0.19$ | *$0.54 \pm 0.072$* | *$0.36 \pm 0.047$* | 79 | 86 |

and Mueller-matrix invariants characterizing the birefringences of myocardial fibrillar network deceased due to mechanical asphyxiation and heart attack are investigated.

2. The method of spatial-frequency filtering of microscopic images of histological sections of biological tissues in the Fourier plane of a polarization-inhomogeneous object field under conditions of azimuthally invariant polarization and Mueller-matrix mapping was tested for the first time and the relationships between the values of statistical moments of the third–fourth orders characterizing polarizing and Mueller-matrix distributions of histological sections of the myocardium and changes in birefringence of polycrystalline networks which are caused by necrotic changes. The growth of the balanced accuracy ($Ac = 80$–$86\%$) of the Fourier filtering method of microscopic and Mueller-matrix images of histological sections of the myocardium in the diagnosis of the cause of death was determined.
3. Based on the polarization-correlation approach to the description of object laser fields formed by spatially ordered and disordered optically anisotropic networks of biological crystals, a comparative study of the diagnostic effectiveness of statistical analysis methods for CDMP mapping of the azimuth distributions and elliptic polarization, as well as CDMP mapping distributions of directional of the optical axes and phase shifts, introduced by birefringent fibrils, was carried out for the first time. The objective parameters most sensitive to changes in optical anisotropy are established—statistical moments of the third and fourth orders characterizing the distribution of the values of the CDMP modulus of microscopic images of biological layers and statistical moments of the first–fourth orders of magnitude characterizing the distribution of the values of the CDMP modulus of polycrystalline histological networks sections of biological tissues
4. The statistical (statistical moments of the first–fourth orders) structure of the distributions of the values of the CDMP modulus of microscopic images of spatially structured fibrillar networks of histological sections of the myocardium that died due to mechanical asphyxiation and heart attack was experimentally studied. The most sensitive parameters were revealed—statistical moments of the first and second orders $Q_1$, $Q_2$ characterizing the coordinate distribution of the values of the CDMP modulus. At the same time, a level of balanced accuracy was achieved—$Ac(V) = 79$–$85\%$ which is higher than the accuracy of azimuthally invariant polarization and Mueller-matrix mapping of biological layers and their microscopic images.
5. The statistical (statistical moments of the first–fourth orders of magnitude) distribution structure of the values of the CDMP modulus of microscopic images of spatially structured fibrillar networks of histological sections of the myocardium deceased due to mechanical asphyxiation and heart attack was experimentally studied. The most sensitive parameters were revealed—a complete set of four statistical moments $Q_{i=1;2;3;4}$ characterizing the coordinate distribution of the values of the CDMA modulus. At the same time, the highest level of balanced accuracy was achieved—$Ac(W) = 84$–$90\%$.

6. For the first time, a scale-selective wavelet analysis of the coordinate distributions of the values of the CDMP and CDMA modulus of microscopic images of optically anisotropic histological sections of biological tissues with structured birefringent fibrillar networks was proposed, and the sensitivity of a set of statistical moments of the second—fourth orders characterizing the distribution of magnitudes of wavelet-coefficient amplitudes was established at different scales of the MHAT function to changes of the distributions of the values of the CDMP and CDMA modulus, which are caused by necrotic changes and achieved a good level of balanced accuracy in the differentiation of such changes at the level of $Ac = 82–85\%$.
7. The method of scale-selective polarization-correlation mapping using spatial-frequency filtering of the coordinate distributions of the values of the CDMP and CDMA modulus of microscopic images of optically anisotropic histological sections of biological tissues with structured birefringent fibrillar networks was developed for the first time and the relationships between the values of the statistical moments of the second–fourth orders that characterize the coordinate distribution of the values of the modulus CDMP and CDMA and the distribution of birefringence of such samples. An excellent level of balanced accuracy $(Ac \sim 90\%)$ of the polarization-correlation method with Fourier filtering of microscopic images of histological sections of the myocardium that died due to mechanical asphyxiation and heart attack was achieved.

# Main Results and Conclusions

1. For the first time, a scale-selective wavelet analysis was used to analyze the azimuthally invariant distributions of the azimuth and polarization ellipticities of microscopic Mueller-matrix images of histological sections of biological tissues with structured optically anisotropic fibrillar networks. The diagnostic capabilities of the wavelet analysis of two-dimensional polarization maps and Mueller-matrix invariants characterizing the birefringences of myocardial fibrillar network deceased due to mechanical asphyxiation and heart attack are investigated.
2. The method of spatial-frequency filtering of microscopic images of histological sections of biological tissues in the Fourier plane of a polarization-inhomogeneous object field under conditions of azimuthally invariant polarization and Mueller-matrix mapping was tested for the first time and the relationships between the values of statistical moments of the third–fourth orders characterizing polarizing and Mueller-matrix distributions of histological sections of the myocardium and changes in birefringence of polycrystalline networks which are caused by necrotic changes. The growth of the balanced accuracy ($Ac = 80$–$86\%$) of the Fourier filtering method of microscopic and Mueller-matrix images of histological sections of the myocardium in the diagnosis of the cause of death was determined.
3. Based on the polarization-correlation approach to the description of object laser fields formed by spatially ordered and disordered optically anisotropic networks of biological crystals, a comparative study of the diagnostic effectiveness of statistical analysis methods for CDMP mapping of the azimuth distributions and elliptic polarization, as well as CDMP mapping distributions of directional of the optical axes and phase shifts, introduced by birefringent fibrils, was carried out for the first time. The objective parameters most sensitive to changes in optical anisotropy are established—statistical moments of the third and fourth orders characterizing the distribution of the values of the CDMP modulus of

 
I. Meglinski et al., *Shedding the Polarized Light on Biological Tissues*,
SpringerBriefs in Applied Sciences and Technology,
https://doi.org/10.1007/978-981-10-4047-4

microscopic images of biological layers and statistical moments of the first–fourth orders of magnitude characterizing the distribution of the values of the CDMP modulus of polycrystalline histological networks sections of biological tissues.

4. The statistical (statistical moments of the first–fourth orders) structure of the distributions of the values of the CDMP modulus of microscopic images of spatially structured fibrillar networks of histological sections of the myocardium that died due to mechanical asphyxiation and heart attack was experimentally studied. The most sensitive parameters were revealed—statistical moments of the first and second orders $Q_1$, $Q_2$ characterizing the coordinate distribution of the values of the CDMP modulus. At the same time, a level of balanced accuracy was achieved—$Ac(V) = 79$–$85\%$—which is higher than the accuracy of azimuthally invariant polarization and Mueller-matrix mapping of biological layers and their microscopic images.
5. The statistical (statistical moments of the first–fourth orders of magnitude) distribution structure of the values of the CDMP modulus of microscopic images of spatially structured fibrillar networks of histological sections of the myocardium deceased due to mechanical asphyxiation and heart attack was experimentally studied. The most sensitive parameters were revealed—a complete set of four statistical moments $Q_{i=1;2;3;4}$ characterizing the coordinate distribution of the values of the CDMA modulus. At the same time, the highest level of balanced accuracy was achieved—$Ac(W) = 84$–$90\%$.
6. For the first time, a scale-selective wavelet analysis of the coordinate distributions of the values of the CDMP and CDMA modulus of microscopic images of optically anisotropic histological sections of biological tissues with structured birefringent fibrillar networks was proposed, and the sensitivity of a set of statistical moments of the second–fourth orders characterizing the distribution of magnitudes of wavelet-coefficient amplitudes was established at different scales of the MHAT function to changes of the distributions of the values of the CDMP and CDMA modulus, which are caused by necrotic changes and achieved a good level of balanced accuracy in the differentiation of such changes at the level of $Ac = 82$–$85\%$.
7. The method of scale-selective polarization-correlation mapping using spatial-frequency filtering of the coordinate distributions of the values of the CDMP and CDMA modulus of microscopic images of optically anisotropic histological sections of biological tissues with structured birefringent fibrillar networks was developed for the first time and the relationships between the values of the statistical moments of the second–fourth orders that characterize the coordinate distribution of the values of the modulus CDMP and CDMA and the distribution of birefringence of such samples. An excellent level of balanced accuracy ($Ac \sim 90\%$) of the polarization-correlation method with Fourier filtering of microscopic images of histological sections of the myocardium that died due to mechanical asphyxiation and heart attack was achieved.

Printed in the United States
by Baker & Taylor Publisher Services